Multi-parametric Optimization and Control

Wiley Series in

Operations Research and Management Science

Operations Research and Management Science (ORMS) is a broad, interdisciplinary branch of applied mathematics concerned with improving the quality of decisions and processes and is a major component of the global modern movement towards the use of advanced analytics in industry and scientific research. The *Wiley Series in Operations Research and Management Science* features a broad collection of books that meet the varied needs of researchers, practitioners, policy makers, and students who use or need to improve their use of analytics. Reflecting the wide range of current research within the ORMS community, the Series encompasses application, methodology, and theory and provides coverage of both classical and cutting edge ORMS concepts and developments. Written by recognized international experts in the field, this collection is appropriate for students as well as professionals from private and public sectors including industry, government, and nonprofit organization who are interested in ORMS at a technical level. The Series is comprised of four sections: Analytics; Decision and Risk Analysis; Optimization Models; and Stochastic Models.

Analytics

Yang and Lee • *Healthcare Analytics: From Data to Knowledge to Healthcare Improvement*
Attoh-Okine • *Big Data and Differential Privacy: Analysis Strategies for Railway Track Engineering*

Forthcoming Titles
Kong and Zhang • *Decision Analytics and Optimization in Disease Prevention and Treatment*

Behavioral Research

Donohue, Katok, and Leider • *The Handbook of Behavioral Operations*

Decision and Risk Analysis

Barron • *Game Theory: An Introduction*, Second Edition

Brailsford, Churilov, and Dangerfield • *Discrete-Event Simulation and System Dynamics for Management Decision Making*

Johnson, Keisler, Solak, Turcotte, Bayram, and Drew • *Decision Science for Housing and Community Development: Localized and Evidence-Based Responses to Distressed Housing and Blighted Communities*

Mislick and Nussbaum • *Cost Estimation: Methods and Tools*

Forthcoming Titles

Aleman and Carter • *Healthcare Engineering*

Optimization Models

Ghiani, Laporte, and Musmanno • *Introduction to Logistics Systems Management*, Second Edition

Forthcoming Titles

Tone • *Advances in DEA Theory and Applications: With Examples in Forecasting Models*

Stochastic Models

Ibe • Random Walk and Diffusion Processes

Forthcoming Titles

Matis • *Applied Markov Based Modelling of Random Processes*

Multi-parametric Optimization and Control

Efstratios N. Pistikopoulos
Nikolaos A. Diangelakis
Richard Oberdieck

This edition first published 2021

Registered Office
John Wiley & Sons, Inc., 111 River Street, Hoboken, NJ 07030, USA

Editorial Office
111 River Street, Hoboken, NJ 07030, USA

For details of our global editorial offices, customer services, and more information about Wiley products visit us at www.wiley.com.

Wiley also publishes its books in a variety of electronic formats and by print-on-demand. Some content that appears in standard print versions of this book may not be available in other formats.

Library of Congress Cataloging-in-Publication Data

Names: Pistikopoulos, Efstratios N., author.
Title: Multi-parametric optimization and control / Efstratios N. Pistikopoulos, Nikolaos A. Diangelakis, Richard Oberdieck.
Description: First edition. | Hoboken, NJ : Wiley, 2021. | Series: Wiley series in operations research and management science | Includes bibliographical references and index.
Identifiers: LCCN 2020024011 (print) | LCCN 2020024012 (ebook) | ISBN 9781119265184 (hardback) | ISBN 9781119265153 (adobe pdf) | ISBN 9781119265191 (epub)
Subjects: LCSH: Mathematical optimization–Computer programs.
Classification: LCC QA402.5 .P558 2021 (print) | LCC QA402.5 (ebook) | DDC 519.7–dc23
LC record available at https://lccn.loc.gov/2020024011
LC ebook record available at https://lccn.loc.gov/2020024012

Cover Design: Wiley
Cover Image: Courtesy of Professor Pistikopoulos'research group

Set in 9.5/12.5pt STIXTwoText by SPi Global, Chennai, India

Printed in United States of America

SKY10021861_102120

To the Memory and Legacy of Professor Christodoulos A. Floudas

Contents

Short Bios of the Authors

Efstratios N. Pistikopoulos

Professor Pistikopoulos is the Director of the Texas A&M Energy Institute and a TEES Eminent Professor in the Artie McFerrin Department of Chemical Engineering at Texas A&M University. He was a Professor of Chemical Engineering at Imperial College London, UK (1991–2015), and the Director of its Centre for Process Systems Engineering (2002–2009). He holds a PhD degree from Carnegie Mellon University and he worked with Shell Chemicals in Amsterdam before joining Imperial. He has authored or co-authored over 500 major research publications in the areas of modeling, control and optimization of process, and energy and systems engineering applications, 12 books, and 2 patents. He is a co-founder of Process Systems Enterprise (PSE) Ltd., a Fellow of AIChE and IChemE and the current Editor-in-Chief of Computers & Chemical Engineering. In 2007, he was a co-recipient of the prestigious MacRobert Award from the Royal Academy of Engineering. In 2012, he was the recipient of the Computing in Chemical Engineering Award of CAST/AIChE. He received the title of Doctor Honoris Causa from the University Politehnica of Bucharest in 2014, and from the University of Pannonia in 2015. In 2013, he was elected Fellow of the Royal Academy of Engineering in the United Kingdom.

Nikolaos A. Diangelakis

Dr. Diangelakis is an Optimization Specialist at Octeract Ltd. in London, UK, a massively parallel global optimization software firm. He was a postdoctoral research associate at Texas A&M University and Texas A&M Energy Institute. He holds a PhD and M.Sc. on Advanced Chemical Engineering from Imperial College London and has been a member of the "Multi-parametric Optimization and Control Group" since late 2011. He earned his bachelor degree in 2011 from the National Technical University of Athens (NTUA). His main research interests are on the area of optimal receding horizon strategies for chemical and energy processes while simultaneously optimizing their design. For that purpose, he is investigating novel solution methods for classes of non-linear, robust, and multi-parametric optimization programming problems. He is the main developer of the PARametric Optimization and Control (PAROC) platform and co-developer of the Parametric OPtimization (POP) toolbox. In 2016 he was chosen as one of five participants in the "Distinguished Junior Researcher Seminars" in Northwestern University, organized by Prof. Fengqi You. In 2017 he received the third place in EFCE's "Excellence Award in Recognition of Outstanding PhD Thesis on CAPE." He is the coauthor of 16 peer reviewed articles, 11 conference papers and 3 book chapters.

Richard Oberdieck

Richard Oberdieck is a Technical Account Manager at Gurobi Optimization, LLC, one of the leading mathematical optimization software companies. He obtained a bachelor and MSc degrees from ETH Zurich in Switzerland (2009–1013), before pursuing a PhD in Chemical Engineering at Imperial College London, UK, which he completed in 2017. During is PhD, he discovered fundamental properties of multi-parametric programming problems and implemented them in the Parametric Optimization (POP) toolbox, of which he was the main developer. After using his knowledge in mathematical modeling and optimization in the space of renewable energies at the world leader in offshore wind energy, Ørsted A/S, he is now helping companies around the world unlock business value through mathematical optimization as a Technical Account Manager for Gurobi Optimization, LLC. He has published 21 papers and 2 book chapters, has an h-index of 11 and was awarded the FICO Decisions Award 2019 in Optimization, Machine Learning and AI.

Preface

Many optimization problems involve parameters that are unknown, either because they cannot be measured, or because they represent information about the future (e.g. future state of a system, future disturbance, future demand). Multi-parametric programming is a technique for the solution of such class of uncertain optimization problems. Through multi-parametric programming, one can obtain the optimization variables of the problem as a function of the bounded uncertain parameters, and the regions (in the space of the parameters) where these functions are valid.

Theoretic and algorithmic developments on multi-parametric programming, along with applications in the area of process systems engineering, have been constantly emerging during the last 30 years.

A variety of algorithms for the solution of a range of classes of multi-parametric programming problems have been developed, with our group publishing over 80 manuscripts, 21 books and book chapters, and 2 patents on the subject. We have further developed a MATLAB© based toolbox, POP©, for the solution of various classes of multi-parametric programming and a framework, PAROC©, for the development of explicit model predictive controllers.

This book aims to enable fundamental understanding in the areas of multi-parametric optimization and control. We hope that by the end of the book, the reader will be able to not only understand almost all aspects of multi-parametric programming, but also judge the key characteristics and particulars of the various techniques developed for different mathematical programming problems, use the tools to solve parametric problems, and finally, develop explicit model predictive controllers.

The book begins with an introduction to the fundamentals of optimization and the basic theories and definitions used in multi-parametric optimization. Then, two main parts follow, providing a clear distinction between algorithmic developments and their applications in the development of explicit model predictive controllers.

Part I focuses on the algorithmic developments of multi-parametric programming. It begins with an overview of the basic sensitivity theorem and progresses to describe solution strategies for linear, quadratic, and mixed-integer multi-parametric problems. A chapter for the solution of the aforementioned classes of problems in MATLAB© is also included. Part I concludes with developments of multi-parametric programming for the solution of other classes of problems.

Part II focuses on multi-parametric model predictive control and its extension for the solution of other control problems. This section concludes with the presentation and applications of PAROC© framework, a framework for the development and closed loop validation of multi-parametric model predictive controllers.

As this book is the outcome of the research work carried out over the last 30 years at the Centre for Process Systems Engineering of Imperial College London and the Texas A&M Energy Institute of Texas A&M University, many colleagues, former and current PhD students, and post-doctorate/research associates have been involved in the presented work. While a number of them are involved in this project as co-authors, we would like to take the opportunity to thank in particular our current research team at Texas A&M Energy Institute, particularly Dr. Styliani Avraamidou and Mr. Iosif Pappas.

We would also like to gratefully acknowledge the financial support kindly provided by our many sponsors, EPSRC, NSF, EU/ERC, DOE/CESMII, DOE/RAPID, Shell, Air Products, Eli-Lilly, and BP.

Finally, we would like to thank Wiley-VCH for their enthusiastic support of this effort.

Richard Oberdieck
College Station, October 2019

Efstratios N. Pistikopoulos
Nikolaos A. Diangelakis

1

Introduction

In this chapter, the fundamental concepts of mathematical optimization and multi-parametric programming will be presented. Such concepts will be the foundation towards the development of state-of-the-art multi-parametric programming strategies and applications, which will appear in this book in the next chapters.

1.1 Concepts of Optimization

1.1.1 Convex Analysis

This section presents the idea of convex sets and introduces function convexity. Convexity plays a vital role to establish the required properties which will enable a multi-parametric solution to hold. In this setting, the following definitions are established.

Definition 1.1 (Line). Consider the points x_1 and $x_2 \in \mathbb{R}^n$. Then the line that passes through these points is defined as

$$\{x | x = (1 - \gamma)x_1 + \gamma x_2, \forall \gamma \in \mathbb{R}\}. \tag{1.1}$$

Definition 1.2 (Line Segment). The closed line segment joining the points $x_1, x_2 \in \mathbb{R}^n$ is defined as:

$$\{x | x = (1 - \gamma)x_1 + \gamma x_2, 0 \leq \gamma \leq 1\}. \tag{1.2}$$

Multi-parametric Optimization and Control, First Edition.
Efstratios N. Pistikopoulos, Nikolaos A. Diangelakis, and Richard Oberdieck.

Definition 1.3 (Convex Set). A set $C \in \mathbb{R}^n$ is a convex set, if the closed line segment joining any two points in the set C belongs to the set C for each γ such that $0 \leq \gamma \leq 1$.

1.1.1.1 Properties of Convex Sets

Let C_1 and C_2 be convex sets defined in $\mathbb{R}^n$. Then

(1) The intersection of $C_1 \cap C_2$ is a convex set.
(2) The summation $C_1 + C_2$ of two convex sets is a convex set.
(3) Let α be a real number. The product αC_1 is a convex set.

Examples of convex sets include lines, polytopes and polyhedra, and open and closed halfspaces.

Definition 1.4 (Convex Function). Let $C \in \mathbb{R}^n$ be a convex subset, and the real function $f(x)$ defined in C. The function $f(x)$ is a convex function if for any x_1,$x_2 \in C$,

$$f\left[(1-\gamma)x_1 + \gamma x_2\right] \leq (1-\gamma)f(x_1) + \gamma f(x_2), 0 \leq \gamma \leq 1. \quad (1.3)$$

If strict inequality holds in expression (1.3) for $0 < \gamma < 1$, then $f(x)$ is a strictly convex function.

Definition 1.5 (Concave Function). Let $C \in \mathbb{R}^n$ be a convex subset, and the real function $f(x)$ defined in C. The function $f(x)$ is a concave function if for any x_1,$x_2 \in C$,

$$f\left[(1-\gamma)x_1 + \gamma x_2\right] \geq (1-\gamma)f(x_1) + \gamma f(x_2), 0 \leq \gamma \leq 1. \quad (1.4)$$

If strict inequality holds in expression (1.4) for $0 < \gamma < 1$, then $f(x)$ is a strictly concave function.

1.1.1.2 Properties of Convex Functions

(1) Let $f_1(x), \ldots, f_n(x)$ be convex functions defined on a convex subset C. Their summation

$$f_1(x) + \cdots + f_n(x) \quad (1.5)$$

is convex, and if at least of one $f_i(x)$ is a strictly convex function, then their summation is strictly convex.

(2) Let a γ be a positive number and $f(x)$ be a (strictly) convex function defined in a convex subset $C \in \mathbb{R}^n$. Then the product $\gamma f(x)$ is (strictly) convex.

(3) Let $f(x)$ be a (strictly) convex function defined in $C \in \mathbb{R}^n$, and $g(y)$ be an increasing convex function defined on the range of $f(x)$ in $\mathbb{R}$. Then, the composite function $g[f(x)]$ defined in C is a (strictly) convex function.

(4) Let $f_1(x), \dots, f_n(x)$ be convex functions defined on a convex subset C. If these functions are bounded from above, their pointwise supremum

$$f(x) = \max\{f_1(x), \dots, f_n(x)\} \tag{1.6}$$

is a convex function on C.

(5) Let $f_1(x), \dots, f_n(x)$ be concave functions defined on a convex subset C. If these functions are bounded from below, their pointwise infimum

$$f(x) = \min\{f_1(x), \dots, f_n(x)\} \tag{1.7}$$

is a concave function on C.

1.1.2 Optimality Conditions

We introduce the following definitions for the solution of general nonlinear optimization problems:

Definition 1.6 (Local Minimum). $x^* \in \mathbb{R}^n$ is called a local minimum if there exists ball of radius ϵ around x^*, $B(x^*)$, such that

$$f(x^*) \leq f(x), \quad \forall x \in B(x^*). \tag{1.8}$$

Definition 1.7 (Global Minimum). $x^* \in \mathbb{R}^n$ is called a global minimum if

$$f(x^*) \leq f(x), \quad \forall x \in \mathbb{R}^n. \tag{1.9}$$

A constrained nonlinear optimization problem, which aims to minimize a real valued function $f(x)$ subject to the inequality constraints $g(x) = \{g_i(x) \leq 0, i \in \mathbb{I}\}$ and equality constraints

$h(x) = \{h_j(x) = 0, j \in \mathbb{J}\}$ is denoted as

$$\begin{aligned} \underset{x}{\text{minimize}} \ & f(x) \\ \text{subject to} \ & g(x) \le 0 \\ & h(x) = 0 \cdot \\ & x \in \mathbb{R}^n \end{aligned} \tag{1.10}$$

Problem (1.10) is a nonlinear optimization problem, if and only if, at least one of $f(x), g_i(x), h_j(x)$ is a nonlinear function. We assume that the aforementioned functions are continuous and differentiable.

Definition 1.8 (Active Constraints). An inequality constraint $g_i(x)$ is called active at a point $\overline{x} \in X$ if $g_i(\overline{x}) = 0$. Conversely, $g_i(x)$ is called inactive if $g_i(\overline{x}) < 0$.

Remark 1.1 If one step of the dual simplex algorithm consists of changing one element of the active set, i.e. let $k_1 = \{i_1, \dots, i_{n-1}, i_n\}$, then the dual pivot involving the constraint i_n yields $k_2 = \{i_1, \dots, i_{n-1}, i_{n+1}\}$.

The first-order constraint qualifications that will be presented in the following text are necessary prerequisites to identify whether a feasible point $\overline{x}$ is a local optimum of the function $f(x)$.

- **Linear independence constraint qualification**: The gradients $\nabla g_j(\overline{x})$ for all $i \in I$ and $\nabla h_i(\overline{x})$ for all $j \in J$ are linearly independent.
- **Slater constraint qualification**: The constraints $g_i(\overline{x})$ for all $i \in I$ are pseudo-convex[1] at $\overline{x}$, while the constraints $h_j(\overline{x})$ for all $j \in J$ are quasi-convex or quasi-concave.[2] In addition, the gradients $\nabla h_j(\overline{x})$ are linearly independent and there exists $\tilde{x}$ such that $g_i(\tilde{x}) < 0$ and $h_j(\tilde{x}) = 0$.

1 A function $f(x)$ is called pseudo-convex if for all feasible x, y where $\nabla f(x)(x - y) \ge 0$ we have $f(y) \ge f(x)$.
2 A function $f(x)$ is called quasi-convex if for all feasible x, y and $\gamma \in [0, 1]$ we have $f(\gamma x + (1 - \gamma)y) \le \max\{f(x), f(y)\}$. Note that a quasi-concave function is a function whose negative is quasi-convex.

1.1.2.1 Karush–Kuhn–Tucker Necessary Optimality Conditions

Let $f(x)$ and $g(x)$ be differentiable at a feasible solution $x^* \in X$, and let $h(x)$ have continuous partial derivatives at x^*. In addition, let p be the number of active inequality constraints at x^*. Then if one of the aforementioned constraint qualifications hold, there exist Lagrange multipliers λ, μ such that

$$\begin{aligned} \nabla_x f(x^*) + \mu^T \nabla_x h(x^*) + \lambda^T \nabla_x g(x^*) &= 0 \\ h(x^*) &= 0 \\ g(x^*) &\leq 0 \\ \lambda_i g_i(x^*) &= 0 \;\; i = 1, 2, \ldots, p \\ \lambda_i &\geq 0 \;\; i = i, 2, \ldots, p \end{aligned} \quad . \tag{1.11}$$

These conditions are the Karush–Kuhn–Tucker (KKT) Necessary Conditions and they are the basis for the solution of nonlinear optimization problems.

1.1.2.2 Karun–Kush–Tucker First-Order Sufficient Optimality Conditions

Consider the sets $J^+ = j : \mu_j > 0$ and $J^- = j : \mu_j < 0$. Then, if the following conditions hold:

- $f(x)$ is pseudo-convex at $\overline{x}$ with respect to all other feasible points x.
- $g_i(x)$ for all $j \in J$ are quasi-convex at $\overline{x}$ with respect to all other feasible points x.
- $h_j(x)$ for all $i \in J^+$ are quasi-convex at $\overline{x}$ with respect to all other feasible points x.
- $h_j(x)$ for all $i \in J^-$ are quasi-concave at $\overline{x}$ with respect to all other feasible points x.

then $\overline{x}$ is a global optimum of problem (1.10). If the aforementioned conditions hold only within a ball of radius ϵ around $\overline{x}$, then $\overline{x}$ is a local optimum of problem (1.10).

1.1.3 Interpretation of Lagrange Multipliers

Consider the following problem:

$$\begin{aligned} \underset{x}{\text{minimize}} \quad & f(x) \\ \text{subject to} \quad & h(x) = \beta \\ & x \in \mathbb{R}^n \end{aligned} \tag{1.12}$$

Let x^* be the global minimum of problem (1.12), and that the gradient of the equality constraints are linearly independent. In addition, assume that the corresponding Lagrange multiplier is λ^*. The vector $\beta = (\beta_1, \beta_2, \dots, \beta_m)$ is a perturbation vector. The solution of problem (1.12) is a function of the perturbation vector along with the multiplier. Hence, the Lagrange function can be written as

$$L(x, \lambda) = f(x(\beta)) + \lambda(\beta)^T[h(x(\beta)) - \beta]. \tag{1.13}$$

Calculating the partial derivative of the Lagrange function with respect to the perturbation vector, we have

$$\nabla_\beta L = \left[\frac{\partial x}{\partial \beta}\right]^T \left(\nabla_x f + \left[\frac{\partial x}{\partial b}\right]^T \lambda\right) + \left[\frac{\partial x}{\partial \beta}\right]^T [h(x)) - \beta] - \lambda \tag{1.14}$$

which yields

$$\nabla_\beta L(x^*, \lambda^*) = -\lambda^* \tag{1.15}$$

Hence, the Lagrange multipliers can be interpreted as a measure of sensitivity of the objective function with respect to the perturbation vector of the constraints at the optimum point x^*.

1.2 Concepts of Multi-parametric Programming

1.2.1 Basic Sensitivity Theorem

Having the essentials of optimization for the purposes of this book covered, the objective of this subchapter is to introduce the role of parameters in an optimization formulation. In this context, the following multi-parametric programming problem is considered:

$$\begin{aligned} \underset{x}{\text{minimize}} \quad & f(x,\theta) \\ \text{subject to} \quad & g(x,\theta) \leq 0 \\ & h(x,\theta) = 0 \\ & x \in \mathbb{R}^n \\ & \theta \in \mathbb{R}^m \end{aligned} \tag{1.16}$$

where x is the vector of the continuous optimization variables, θ is the vector of the uncertain parameters, and the sets $i \in \mathbb{I}, j \in \mathbb{J}$ correspond to the inequality and equality constraint sets, respectively.

Theorem 1.1 ***(Basic Sensitivity Theorem, [1])*** *Let a general multi-parametric programming problem be described by (1.16). Assume that the functions defining problem (1.16) are twice differentiable in x and their gradients with respect to x and the constraints are once continuously differentiable in θ in a neighborhood of (x^*, θ^*). In addition, assume that the second-order sufficient conditions for a local minimum of the problem hold at x^* with associated Lagrange multipliers λ^* and μ^*. Lastly, let the gradients $\nabla g_i(x^*, \theta^*)$ (for $i \in \mathbb{I}$ such that $g_i(x^*, \theta^*) = 0$) and $\nabla h_j(x^*, \theta^*)$ be linearly independent (i.e. LICQ holds), and $\lambda_i \geq 0$ for $i \in \mathbb{I}$ such that $g_i(x^*, \theta^*) = 0$, i.e. strict complementary slackness (SCS) holds.*

Then, the first-order sensitivity results for a second-order local minimizing point x^* are known as the basic sensitivity theorem (BST), and the following properties hold:

- *x^* is a local isolated minimizing point of the problem and the associated Lagrange multipliers λ_i^* and μ_j^* are unique.*
- *For θ in the neighborhood or θ^*, there exists a unique, but continuously differentiable vector function $\eta(\theta) = [x(\theta), \lambda(\theta), \mu(\theta)]^T$ satisfying the second-order sufficient conditions for a local minimum of the problem with associated unique Lagrange multipliers $\lambda(\theta)$ and $\mu(\theta)$.*
- *For θ near θ^* the set of binding inequalities is unchanged, SCS holds and the binding constraint gradients are linearly independent at $x(\theta)$.*

Proof: See [1]. □

If there exist Lagrange multipliers, λ_i^* and μ_j^*, such that the first-order KKT conditions hold, then we have:

$$\begin{aligned} L(x^*, \mu^*, \lambda^*, \theta^*) &= \nabla_x f(x^*, \theta^*) + \mu^T \nabla_x h(x^*, \theta^*) + \lambda^T \nabla_x g(x^*, \theta^*) = 0 \\ h(x^*, \theta^*) &= 0 \\ g(x^*, \theta^*) &\leq 0 \\ \lambda_i g_i(x^*, \theta^*) &= 0, \quad i = 1, 2, \ldots, p \\ \lambda_i &\geq 0, \quad i = i, 2, \ldots, p \end{aligned} \tag{1.17}$$

and the vector $F(x, \lambda, \mu, \theta)$ is defined as follows:

$$F(x, \lambda, \mu, \theta) = \begin{bmatrix} \nabla_x L(x, \lambda, \mu, \theta) \\ \lambda_i g_i(x, \theta) \\ h(x, \theta) \end{bmatrix} = 0 \tag{1.18}$$

Furthermore, if there exists $z(x)$ for which

$$z(x) \nabla_{xx} L(\eta, \theta) z(x) \geq 0, \quad \forall z \neq 0 \tag{1.19}$$

the Basic Sensitivity Theorem holds, and it is identically satisfied for a neighborhood θ around θ^* and can be differentiated with respect to θ to yield explicit expressions for the partial derivatives of the vector function $\eta(\theta) = [x(\theta), \lambda(\theta), \mu(\theta)]^T$.

The first-order estimate of the variation of an isolated local solution $x(\theta)$ of (1.16) and the associated unique Lagrange multipliers $\lambda(\theta)$ and $\mu(\theta)$ can be approximated, given that $\eta(\theta^*) = [x(\theta^*), \lambda(\theta^*), \mu(\theta^*)]^T$ is known and that $\nabla_\theta \eta(\theta^*)$ is available.

In particular, let a be the concatenation of the vectors η and θ $a = [\eta^T | \theta^T]^T$. The first-order Taylor expansion of the vector F around a^* can be expressed as follows:

$$F(a) = \nabla_a F(a^*)(a - a^*) + F(a^*) \tag{1.20}$$

Under the assumptions and the principles of the Basic Sensitivity Theorem, in a neighborhood of a^* the first-order KKT conditions hold and the value of F(a) around a^* remains zero. For systems that consist of polynomial objective functions of up to second degree and linear constraints, with respect to the optimization variables and the uncertain parameters, the first-order Taylor expansion is exact. Hence, the exact multi-parametric solution can be obtained for the following multi-parametric quadratic programming problem

$$\begin{aligned}
\min_{x} \quad & \tfrac{1}{2}x^T Q x + x^T H^T \theta + c_x^T x \\
\text{s.t.} \quad & A_i x \leq b_i + F_i \theta \\
& A_j x = b_j + F_j \theta \\
& \theta \in \Theta := \{\theta \in \mathbb{R}^m \mid CR_A \theta \leq CR_b\} \\
& x \in \mathbb{R}^n
\end{aligned}, \tag{1.21}$$

where matrices $A_{i(1\times n)}$, and $F_{i(1\times m)}$, $A_{j(1\times n)}$, $F_{j(1\times m)}$ and the scalars b_i, b_j correspond to the ith and jth inequality and equality constraints of the sets $\mathbb{I}$ and $\mathbb{J}$, respectively. This problem serves as the basis that will be discussed in Part I, where its solution properties and solution strategies among other things are in focus. Part II then focusses on the application of such problems to optimal control, as the use of parameters enables the formulation of explicit model predictive control problems.

1.3 Polytopes

Multi-parametric programming is intimately related to the properties and operations applicable to polytopes. In the following, some basic definitions on polytopes are stated, which are used throughout the book.

Definition 1.9 A function $x(\theta) : \Theta \mapsto \mathbb{R}^n$, where $\Theta \in \mathbb{R}^q$ is a polytope, is called piecewise affine if it is possible to partition Θ into disjoint polytopes, called critical regions, CR_i and

$$x(\theta) = K_i \theta + r_i, \ \forall \theta \in CR_i. \tag{1.22}$$

Remark 1.2 The definition of piecewise quadratic is analogous.

Definition 1.10 The set $\mathcal{P}$ is called a n-dimensional polytope if it satisfies

$$\mathcal{P} := \left\{x \in \mathbb{R}^n | a_i^T x \leq b_i, i = 1, \ldots, m\right\}, \tag{1.23}$$

where m is finite.

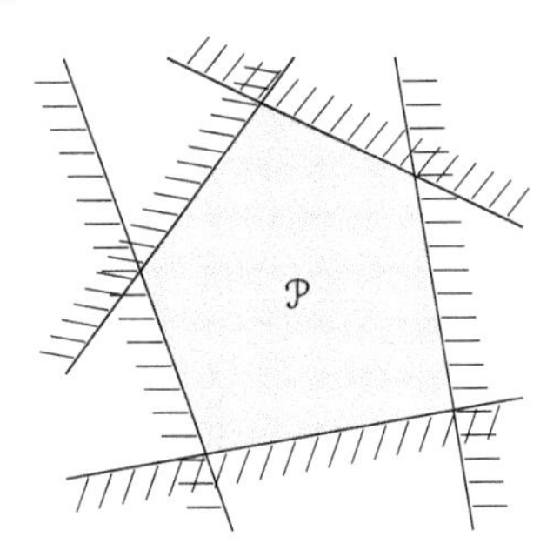

Figure 1.1 A schematic representation of a two-dimensional polytope $\mathcal{P}$.

A schematic representation of a polytope is given in Figure 1.1.
In addition to Definition (1.10), the following well-known characteristics of polytopes are considered:

- A polytope is called bounded if and only if there exists a finite $x_{\min} \in \mathbb{R}^n$ and $x_{\max} \in \mathbb{R}^n$ such $x_{\min} \leq x \leq x_{\max}$ for all $x \in \mathcal{P}$.
- A polytope, which is closed and bounded, is called compact.
- Let $\mathcal{P}$ be an n-dimensional polytope. Then, a subset of a polytope is called a face of $\mathcal{P}$ if it can be represented as

$$\mathcal{F} = \mathcal{P} \cap \left\{x \in \mathbb{R}^n | a^T x = b\right\} \tag{1.24}$$

  for some inequality $a^T x \leq b$, which holds for all $x \in \mathcal{P}$. The faces of polytopes of dimension $n-1$, 1, and 0 are referred to as facets, edges, and vertices, respectively.
- Two polytopes $\mathcal{P}_1$ and $\mathcal{P}_2$ are called disjoint if $\text{int}(\mathcal{P}_1) \cap \text{int}(\mathcal{P}_2) = \emptyset$. Similarly, two polytopes $\mathcal{P}_1$ and $\mathcal{P}_2$ are called overlapping if $\text{int}(\mathcal{P}_1) \cap \text{int}(\mathcal{P}_2) \neq \emptyset$. Lastly, two polytopes $\mathcal{P}_1$ and $\mathcal{P}_2$ are called adjacent or neighboring if $\mathcal{P}_1 \cap \mathcal{P}_2$ is a $n-1$-dimensional polytope.
- Let $\mathcal{P}_1$ and $\mathcal{P}_2$ be two adjacent polytopes. Then the facet-to-facet property is said to hold if $F = \mathcal{P}_1 \cap \mathcal{P}_2$ is a facet of both $\mathcal{P}_1$ and $\mathcal{P}_2$ (see Figure 1.2 for an illustration).
- Let $\mathcal{P}$ be an n-dimensional polytope. Then, there exists a series of k vertices $x_i \in \mathbb{R}^n$ such that

$$\mathcal{P} := \left\{x \in \mathbb{R}^n | x = \sum_{i=1}^{k} \lambda_i x_i, \sum_{i=1}^{k} \lambda_i = 1, \lambda_i \geq 0\right\}. \tag{1.25}$$

- Eq. (1.23) is referred to the halfspace (or H) representation, while Eq. (1.25) denotes the vertex (or V) representation. The process

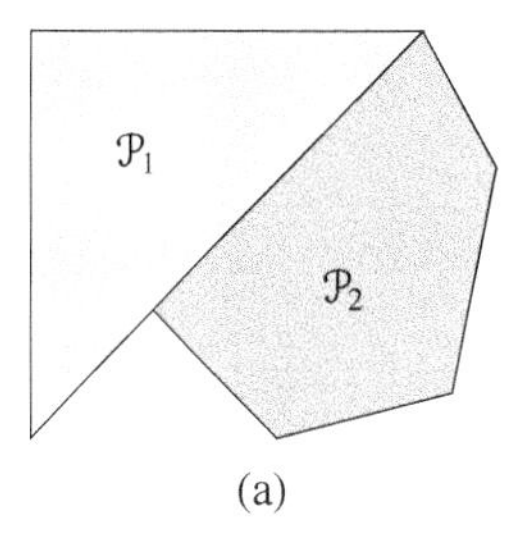

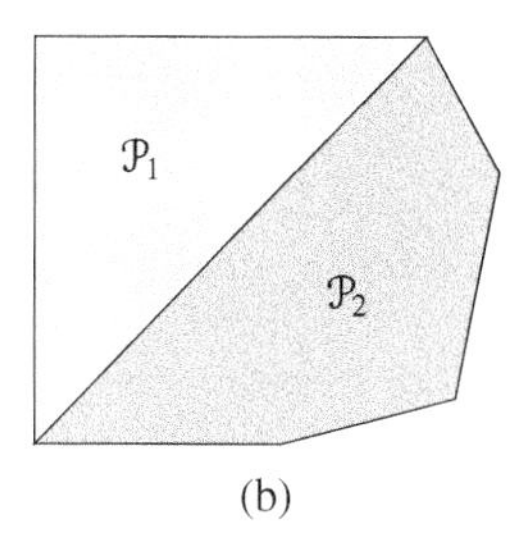

Figure 1.2 A schematic representation of the differences between two polytopes $\mathcal{P}$ and $\mathcal{P}_2$ (a), which are adjacent and (b) where the facet-to-facet property holds. Clearly, while all polytopes that satisfy the facet-to-facet property are adjacent, the opposite may not be true.

of moving from the halfspace to the vertex representation is referred to as vertex enumeration.

- The Chebyshev center of a polytope is given as the largest Euclidean ball that lies in a polytope [2]. It can be determined by solving the following linear programming (LP) problem:

$$\begin{aligned} R = \ & \underset{x,r}{\text{minimize}} \ -r \\ & \text{subject to } A_i x + r||A_i||_2 \leq b_i, \quad \forall i = 1, \ldots, m, \end{aligned} \tag{1.26}$$

  where the solution R denotes the radius of the largest Euclidean ball. Based on the solution of problem (1.26), the following conclusions can be drawn:
  - Problem (1.26) is infeasible: The polytope is empty.
  - $R = 0$: The polytope is lower-dimensional.
  - $R > 0$: The polytope is full-dimensional.

1.3.1 Approaches for the Removal of Redundant Constraints

A concept that is very important in multi-parametric programming is the aspect of redundancy:

Theorem 1.2 *([3]) Consider an n-dimensional compact polytope $\mathcal{P}$ in halfspace representation. A constraint $a_i^T x \leq b_i$ is redundant if and only if*

$$\mathcal{P}_i = \{x \in \mathbb{R}^n | A_i x > b_i, A_k x \leq b_k, \forall k \neq i\} = \emptyset. \tag{1.27}$$

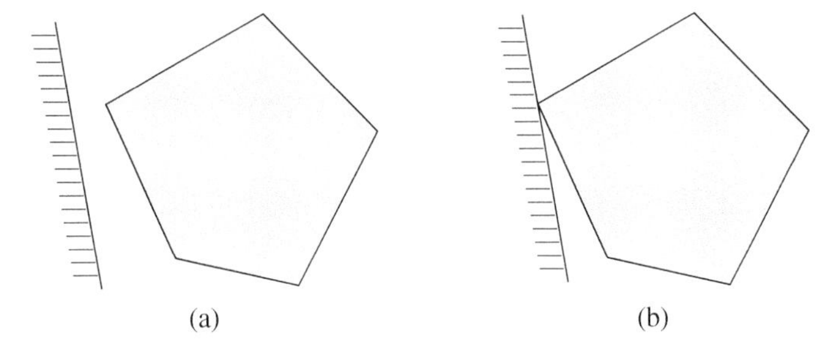

Figure 1.3 A schematic representation of (a) strongly and (b) weakly redundant constraints.

Additionally, a constraint $A_i x \leq b_i$ is strongly redundant if and only if

$$\mathcal{P}_i' = \{x \in \mathbb{R}^n | A_i x \geq b_i, A_k x \leq b_k, \forall k \neq i\} = \emptyset. \tag{1.28}$$

Remark 1.3 A constraint is called weakly redundant if it is redundant but not strongly redundant, i.e. Eq. (1.27) but not Eq. (1.28) holds. A schematic representation of weakly and strongly redundant constraints is shown in Figure 1.3.

If a polytope $\mathcal{P}$ does not feature any redundant constraints, it is said to be in minimal representation.

Consider an n-dimensional compact polytope $\mathcal{P} = \{x \in \mathbb{R}^n | Ax \leq b\}$, where $A \in \mathbb{R}^{m \times n}$ and $b \in \mathbb{R}^m$. The following strategies aim at identifying the minimal representation of $\mathcal{P}$:

Remark 1.4 Here, two of the most common approaches used are reported. The field of the removal of redundant constraints has been widely studied, and its review is beyond the scope of this book. The reader is referred to [3, 4] for an interesting treatment of the matter.

1.3.1.1 Lower-Upper Bound Classification

Given the bounds $l_j \leq x_j \leq u_j, \forall j = 1, \ldots, m$, a constraint $A_i x \leq b_i$ is redundant if

$$U_i \leq b_i, \tag{1.29}$$

where

$$U_i = \sum_{j \in P_j} A_{ij} u_j + \sum_{j \in N_j} A_{ij} l_j, \tag{1.30}$$

where $P_j = \{j | A_{ij} > 0\}$ and $N_j = \{j | A_{ij} < 0\}$. This approach relies on the identification of the worst-case scenario given the lower and upper bounds [5]. If these bounds are not available, they can be calculated by solving the following $2n$ LP problems [6]:

$$\begin{aligned} &\underset{x}{\text{Minimize}} \ \pm x_i \\ &\text{Subject to } x \in \mathcal{P}. \end{aligned} \tag{1.31}$$

1.3.1.2 Solution of Linear Programming Problem

Consider the following constraint-specific version of problem (1.26):

$$\begin{aligned} R_i = \underset{x,r}{\text{minimize}} \quad & -r \\ \text{subject to} \quad & Ax \leq (b - ||A^i||_2 r) \\ & A_i x = b_i \\ & ||A^i||_2 = ||1 - (AA_i^T)^2||_2 \\ & x \in \mathcal{P}, \ r \in \mathbb{R}, \end{aligned} \tag{1.32}$$

where $(\cdot)^2$ denotes the element-wise square of $(\cdot)$. Note that $Ax \leq b$ is assumed to be normalized such that $||a_i||_2 = 1$ for all $i = 1, \ldots, m$. Then the ith constraint is redundant if and only if $R_i \leq 0$. Note that this identifies weakly and strongly redundant constraints.

Remark 1.5 The solution of problem (1.32) identifies the largest Euclidean ball which on the set $\mathcal{K} = \{x | x \in \mathcal{P} \cup A_i x = b_i\}$, i.e. which lies on the ith constraint. Thus, the solution can be understood as the center of the ith constraint with respect to $\mathcal{P}$.

1.3.2 Projections

One of the operations used in this book is the (orthogonal) projection:

Definition 1.11 (Projection [7]).Let $P \subset \mathbb{R}^d \times \mathbb{R}^k$ be a polytope. Then the projection $\pi_d(P)$ of P onto $\mathbb{R}^d$ is defined as:

$$\pi_d(P) = \left\{x \in \mathbb{R}^d | \exists y \in \mathbb{R}^k, (x, y) \in P\right\}. \tag{1.33}$$

Projecting polytopes is one of the fundamental operations in computational geometry and has many applications in control theory. Two commonly encountered strategies for the calculation of the projection are the following:

- Solving a multi-parametric linear programming (mp-LP) problem (see e.g. [8])
- Performing a Fourier–Motzkin (FM) elimination (see, e.g. [9])

In addition, the concept of a hybrid projection is introduced:

Definition 1.12 (Hybrid Projection). Consider the set $P \subset \mathbb{R}^d \times \mathbb{R}^k \times \{0,1\}^r$. Then, the hybrid projection $\tilde{\pi}_d(P)$ of P onto $\mathbb{R}^d$ is defined as

$$\tilde{\pi}_d(P) = \left\{x \in \mathbb{R}^d | \exists y \in \mathbb{R}^k \times \{0,1\}^r, (x,y) \in P\right\}. \tag{1.34}$$

By inspection it is clear that (i) $\tilde{\pi}_d(P)$ is obtained by performing at most 2^r projections, one for each combination of the binary variables and consequently (ii) $\tilde{\pi}_d(P)$ is generally a union of at most 2^r possibly overlapping polytopes.

A hybrid projection can thereby be performed by solving a multi-parametric mixed-integer programming problem purely based on feasibility requirements.

1.3.3 Modeling of the Union of Polytopes

The aim is to represent a union of polytopes $P = \bigcup_{i=1}^{p} \{x | G_i x \leq g_i\}$ as a single set of linear inequality constraints in order to seamlessly include them within multi-parametric programming problems. However, in order to address the possible non-convexity within unions of polytopes, the introduction of suitable binary variables is required. First, consider that a point $x \in P$ if and only if there exists at least one i such that $G_i x \leq g_i$. Thus, one binary variable y_i is defined such that

$$[G_i x \leq g_i] \rightarrow [y_i = 1] \tag{1.35a}$$

$$\sum_{i=1}^{p} y_i \geq 1. \tag{1.35b}$$

Let $G_{i,j}$ and $g_{i,j}$ denote the jth row and element of $G_i \in \mathbb{R}^{t_i \times n}$ and $g_i \in \mathbb{R}^{t_i}$, respectively. Then, the statement $G_i x \leq g_i$ holds if and only if $G_{i,j} x \leq g_{i,j}, \forall j$. Thus, one binary variable per row of G_i, y_i^j is defined such that

$$[G_{i,j} x \leq g_{i,j}] \leftrightarrow [y_i^j = 1] \tag{1.36a}$$

$$\left[\sum_{j=1}^{t_i} y_i^j = t_i\right] \rightarrow [y_i = 1] \tag{1.36b}$$

$$\sum_{i=1}^{p} y_i \geq 1. \tag{1.36c}$$

Based on [10, 11], Eqs. (1.1.36a)–(1.1.36c) are reformulated as

$$G_{i,j}^T x + M y_i^j \leq M + g_{i,j} \tag{1.37a}$$

$$G_{i,j}^T x - m y_i \geq g_{i,j} \tag{1.37b}$$

$$t_i y_i \leq \sum_{j=1}^{t_i} y_i^j \tag{1.37c}$$

$$y_i \geq \sum_{j=1}^{t_i} y_i^j + 1 - t_i, \tag{1.37d}$$

where $m \leq x \leq M, \forall x \in P$. Thus, the final formulation of the union as a set of linear inequality constraints featuring binary variables is given as

$$P = \bigcup_{i=1}^{p} \{x | G_i x \leq g_i\} \rightarrow \left\{ \begin{array}{r l} G_{i,j}^T x + M y_i^j & \leq M + g_{i,j} \\ -G_{i,j}^T x + m y_i^j & \leq -g_{i,j} \\ t_i y_i - \sum_{j=1}^{t_i} y_i^j & \leq 0 \\ -y_i + \sum_{j=1}^{t_i} y_i^j & \leq t_i - 1 \\ -\sum_{i=1}^{p} y_i & \leq -1 \end{array} \right. . \tag{1.38}$$

1.4 Organization of the Book

The remainder of this book is organized in two parts. In the first part, the theoretical and algorithmic essentials of multi-parametric programming problems will be established. These include algorithms for the solution of multi-parametric linear programming (mp-LP), multi-parametric quadratic programming (mp-QP), multi-parametric mixed-integer linear programming (mp-MILP), and multi-parametric mixed-integer quadratic programming (mp-MIQP) problems. On the other hand, the latter of these parts is focused on the applications of multi-parametric programming and specifically on its utilization to provide solutions to receding horizon optimization problems such as model predictive control.

References

1 Fiacco, A.V. (1983) *Introduction to sensitivity and stability analysis in nonlinear programming, Mathematics in science and engineering*, vol. v, 165, Academic Press, New York.
2 Boyd, S.P. and Vandenberghe, L. (2004) *Convex optimization*, Cambridge University Press, Cambridge, UK and New York.
3 Telgen, J. (1981) *Redundancy and linear programs*, Mathematisch Centrum, Amsterdam.
4 Karwan, M.H., Lotfi, V., Telgen, J., and Zionts, S. (1983) *Redundancy in mathematical programming: a state-of-the-art survey, Lecture notes in economics and mathematical systems*, vol. 206, Springer-Verlag, Berlin and New York.
5 Brearley, A.L., Mitra, G., and Williams, H.P. (1975) Analysis of mathematical programming problems prior to applying the simplex algorithm. *Mathematical Programming*, 8 (1), 54–83, doi: 10.1007/BF01580428.
6 Suard, R., Lofberg, J., Grieder, P., Kvasnica, M., and Morari, M. (2004) Efficient computation of controller partitions in multi-parametric programming, in *2004 43rd IEEE Conference on Decision and Control (CDC)*, vol. 4, pp. 3643–3648, doi: 10.1109/CDC.2004.1429297.

7 Jones, C.N., Kerrigan, E.C., and Maciejowski, J.M. (2004) Equality Set Projection: A new algorithm for the projection of polytopes in halfspace representation. Technical Report CUED/F-INFENG/TR.463. Cambridge University, Cambridge, UK.

8 Kouramas, K.I., Panos, C., Faísca, N.P., and Pistikopoulos, E.N. (2013) An algorithm for robust explicit/multi-parametric model predictive control. *Automatica*, 49 (2), 381–389, doi: 10.1016/j.automatica.2012.11.035. URL http://www.sciencedirect.com/science/article/pii/S0005109812005717.

9 Schrijver, A. (1998) *Theory of linear and integer programming*, Wiley-interscience series in discrete mathematics and optimization, Wiley, Chichester and New York.

10 Bemporad, A. and Morari, M. (1999) Control of systems integrating logic, dynamics, and constraints. *Automatica*, 35 (3), 407–427, doi: 10.1016/S0005-1098(98)00178-2. URL http://www.sciencedirect.com/science/article/pii/S0005109898001782.

11 Williams, H.P. (2013) *Model building in mathematical programming*, Wiley, Hoboken, NJ, 5th edn.

Part I

Multi-parametric Optimization

2

Multi-parametric Linear Programming

Consider the following linear programming (LP) problem:

$$
\begin{aligned}
z = \underset{x}{\text{Minimize}} \ & c^T x \\
\text{Subject to} \ & Ax \leq b \\
& A_{\text{eq}} x = b_{\text{eq}} \\
& x \in \mathbb{R}^n,
\end{aligned}
\tag{2.1}
$$

where $c \in \mathbb{R}^n$, $A \in \mathbb{R}^{m \times n}$, $b \in \mathbb{R}^m$, $A_{\text{eq}} \in \mathbb{R}^{s \times n}$, $b_{\text{eq}} \in \mathbb{R}^s$, $Ax \leq b$ is a compact polytope and A_{eq} is assumed to have full rank.[1] This problem formulation requires the deterministic knowledge of all elements of problem (2.1). However, in many applications values such as prices, demand, and risk might change over time and have great impact on the solution. In order to take this into account, the uncertainty can be considered explicitly by formulating a multi-parametric linear programming (mp-LP) problem:

$$
\begin{aligned}
z(\theta) = \underset{x}{\text{Minimize}} \ & c^T x \\
\text{Subject to} \ & Ax \leq b + F\theta \\
& A_{\text{eq}} x = b_{\text{eq}} + F_{\text{eq}} \theta \\
& \theta \in \Theta := \{\theta \in \mathbb{R}^q | \text{CR}_A \theta \leq \text{CR}_b\} \\
& x \in \mathbb{R}^n,
\end{aligned}
\tag{2.2}
$$

where $F \in \mathbb{R}^{m \times q}$, $F_{\text{eq}} \in \mathbb{R}^{s \times q}$, $\text{CR}_A \in \mathbb{R}^{r \times q}$, $\text{CR}_b \in \mathbb{R}^r$ and Θ is a compact polytope. The difference between problems (2.1) and (2.2) is the presence of the bounded uncertain parameters θ in the

1 If A_{eq} does not have full rank, it is always possible to find an equivalent matrix A'_{eq} with a reduced number of rows, which has full rank.

Multi-parametric Optimization and Control, First Edition.
Efstratios N. Pistikopoulos, Nikolaos A. Diangelakis, and Richard Oberdieck.

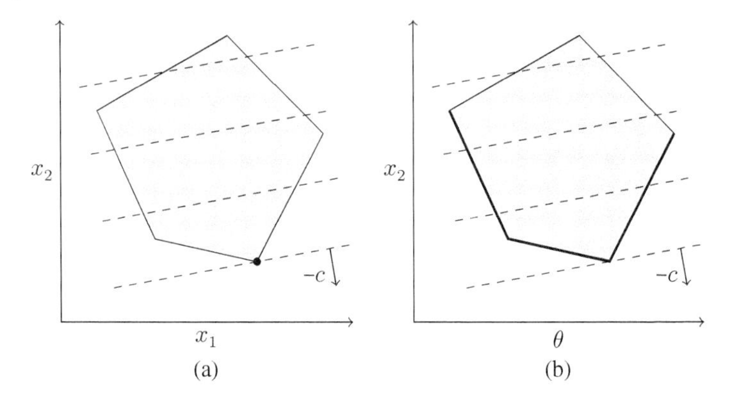

Figure 2.1 The difference between the solution of an LP and an mp-LP problem (black dot and line, respectively), where the mp-LP problem is obtained by treating x_1 as a parameter. In (a), the solution is a single point in one of the vertices of the feasible space, while in (b) the solution is a function of the parameter θ.

constraints. As a result, the solution x, the Lagrange multipliers λ and μ, and the optimal objective function z of problem (2.2) are obtained as a function of θ, i.e. $x(\theta)$, $\lambda(\theta)$, $\mu(\theta)$, and $z(\theta)$, respectively. A schematic representation of the difference between problem (2.1) and (2.2) is shown in Figure 2.1.

Figure 2.2 shows some of the properties of the solution of the mpLP problem (2.2).

Remark 2.1 Note that it is possible to add a scalar d to the objective function of an LP problem without influencing the optimal solution. Similarly, it is possible to add an arbitrary scaling function $f(\theta)$ to an mp-LP problem without influencing the optimal solution.

2.1 Solution Properties

Remark 2.2 Due to the similarities between mp-LP and multi-parametric quadratic programming (mp-QP) problems, the different solution strategies available are discussed in detail in Chapter 4.

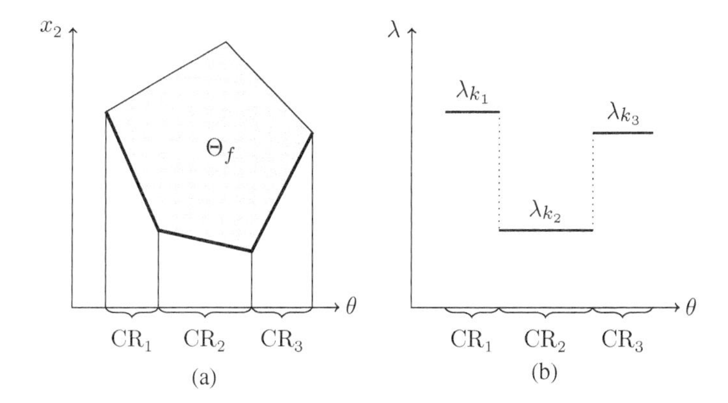

Figure 2.2 A schematic representation of the solution of the mp-LP problem from Figure 2.1. In (a), the partitioning of the convex, feasible parameter space Θ_f, as well as the piecewise affine nature of the optimal solution $x(\theta)$ and the convex and piecewise affine nature of the objective function $z(\theta)$ is shown. In (b), the Lagrange multipliers λ_k as a function of θ are shown.

2.1.1 Local Properties

Consider a fixed, nominal point $\theta = \theta_0$, which transforms the mp-LP problem (2.2) into an LP problem of type (2.1). The solution of this LP problem at θ_0 yields the solution x_0 as well as the values of the Lagrange multipliers λ_0. Based on these, the indices of the active set k are identified:

$$A_k x_0 = b_k + F_k \theta_0 \tag{2.3a}$$

$$\lambda_{0,k} > 0. \tag{2.3b}$$

Remark 2.3 In the case where the set k from Eq. (2.3) is not unique, the solution is said to be degenerate. The impact of degeneracy on the parametric solution is discussed in Chapter 2.2.

Together with the equality constraints, which have to be satisfied for any θ, the following active set matrices and vectors are defined:

$$A_{\mathrm{AS}} = \left[A_k^T \; A_{\mathrm{eq}}^T\right]^T \tag{2.4a}$$

$$b_{AS} = \left[b_k^T \; b_{eq}^T\right]^T \tag{2.4b}$$

$$F_{AS} = \left[F_k^T \; F_{eq}^T\right]^T. \tag{2.4c}$$

Note that A_{AS} has to have full rank in order to fulfill the LICQ condition described in Chapter 1. Since the objective function is linear and the constraints are affine, the change of the solution of problem (2.2) based on the basic sensitivity theorem is given by[2]:

$$x(\theta) = A_{AS}^{-1}(F_{AS}\theta + b_{AS}) \tag{2.5a}$$

$$\lambda(\theta) = \lambda_0 \tag{2.5b}$$

$$\mu(\theta) = \mu_0. \tag{2.5c}$$

Based on Eq. (2.5), the following statements regarding the solution around θ_0 can be made:

- The optimization variables $x(\theta)$ are affine functions of the parameter θ.
- In the case of mp-LP problems, the values of the Lagrange multipliers $\lambda(\theta)$ and $\mu(\theta)$ do not change as a function of θ around a nominal point θ_0.
- The square matrix A_{AS} is invertible since the SCS and LICQ conditions of Chapter 1 have to hold.
- In order for Eq. (2.5) to remain the optimal solution around a nominal point θ_0, it needs to be feasible, i.e.

$$Ax(\theta) \leq b + F\theta \qquad \text{(Feasibility of } x(\theta)\text{)} \tag{2.6a}$$

$$\mathrm{CR}_A\theta \leq \mathrm{CR}_b \qquad \text{(Feasibility of } \theta\text{)} \tag{2.6b}$$

Note that since the values of the Lagrange multipliers λ do not change as a function of θ, the optimality requirement $\lambda(\theta) \geq 0$ from the Karush-Kuhn-Tucker conditions can be omitted from the construction of the feasible region.

Thus, the optimal solution of problem (2.2) around θ_0 is given by Eq. (2.5) and is valid in the compact polytope described by Eq. (2.6),

---

2 Note that this solution can also be directly obtained by solving the set of equations $A_{AS}x = b_{AS} + F_{AS}\theta$ for x, which corresponds to the propagation of the solution of the LP at θ_0 along the parameter space.

which is referred to as *critical region*:

$$\mathrm{CR} = \left\{ \theta \in \mathbb{R}^q \,\middle|\, \begin{bmatrix} A(A_{\mathrm{AS}}^{-1}F_{\mathrm{AS}}) - F \\ \mathrm{CR}_A \end{bmatrix} \theta \leq \begin{bmatrix} b - A(A_{\mathrm{AS}}^{-1}b_{\mathrm{AS}}) \\ \mathrm{CR}_b \end{bmatrix} \right\}. \tag{2.7}$$

Based on Eq. (2.7), the following Lemmata result:

Lemma 2.1 *Every critical region is uniquely defined by its active set.*

Proof: By inspection of Eq. (2.7), it is clear that the differences between any two critical regions are the values of A_k, b_k, and F_k, respectively, which only depend on the active set k. Thus, the set of active constraints k uniquely defines the critical region CR, which completes the proof. □

Lemma 2.2 *The maximum number of critical regions, $\phi_{\max}$, for problem (2.2) is given by*

$$\phi_{\max} = \binom{m}{n-s} = \frac{m!}{(n-s)!(m-(n-s))!}. \tag{2.8}$$

Proof: Consider $\theta = \theta_0$. Then an optimal solution of the resulting LP problem is guaranteed to lie in a vertex, thus featuring n active constraints. However, as the equality constraints have to be fulfilled for all θ, the number of active inequality constraints is given by $n - s$, where s is the number of equality constraints. As the number of critical regions is uniquely defined by the active set, it is bound by above by all possible combinations of active sets, which is given by $\binom{m}{n-s}$, which completes the proof. □

2.1.2 Global Properties

The solution properties described in Chapter 2.1.1 hold for any feasible point θ_0 and thus the following theorem can be formulated:

Theorem 2.1 ***(The Solution of mp-LP Problems)*** *Consider the mp-LP problem (2.2). Then the set of feasible parameters $\Theta_f \subseteq \Theta$ is convex, the optimizer $x(\theta) : \Theta_f \mapsto \mathbb{R}^n$ is continuous and* *piecewise affine, and the optimal objective function $z(\theta) : \Theta_f \mapsto \mathbb{R}$ is continuous, convex, and piecewise affine.*

Proof: The two key statements that need to be proven are the convexity of Θ_f and $z(\theta)$. Consider two generic parameter values $\theta_1, \theta_2 \in \Theta_f$ and let $z(\theta_1)$, $z(\theta_2)$ and $x_1 = x(\theta_1)$ and $x_2 = x(\theta_2)$ be the corresponding optimal objective function values and optimizers. Additionally, let $\alpha \in [0,1]$ and define $x_\alpha = \alpha x_1 + (1-\alpha)x_2$ and $\theta_\alpha = \alpha\theta_1 + (1-\alpha)\theta_2$. Then, since $\theta_1, \theta_2 \in \Theta_f$, x_1, and x_2 are feasible and satisfy the constraints $Ax_1 \leq b + F\theta_1$ and $Ax_2 \leq b + F\theta_2$. As these constraints are affine, they can be linearly combined to obtain $Ax_\alpha \leq b + F\theta_\alpha$, and therefore x_α is feasible for the optimization problem (2.2). Since a feasible solution x_α exists at θ_α, an optimal solution exists at θ_α and thus Θ_f is convex.

The optimal solution at θ_α will be less than or equal to the feasible solution, i.e. $z(\theta_\alpha) \leq c^T x_\alpha$ and thus:

$$
\begin{aligned}
&z(\theta_\alpha) - [\alpha(c^T x_1) + (1-\alpha)(c^T x_2)] \\
&\quad \leq c^T x_\alpha - [\alpha(c^T x_1) + (1-\alpha)(c^T x_2)] \qquad (2.9a) \\
&\quad = 0, \qquad (2.9b)
\end{aligned}
$$

i.e. $z(\alpha\theta_1 + (1-\alpha)\theta_2) \leq \alpha z(\theta_1) + (1-\alpha)z(\theta_2)$, $\forall \theta_1, \theta_2 \in \Theta_f$, $\forall \alpha \in [0,1]$. This proves the convexity of $z(\theta)$ and Θ_f. The piecewise affine nature of $x(\theta)$ and $z(\theta)$ is a direct result from the fact that the boundary between two regions belongs to both regions. Since the optimum is unique, the optimizer and thus the optimal objective function value must be continuous across the boundary. □

In addition to the fundamental properties derived in Theorem 2.1, it is possible to infer more structural information about the connections between the critical regions:

Definition 2.1 (mp-LP Graph). Let each active set k of an mp-LP problem be a node in the set of solutions S. Then the nodes k_1 and k_2 are connected if (i) there exists a $\theta^* \in \Theta_f$ such that k_1 and k_2 are both optimal active sets and (ii) it is possible to pass from k_1 to k_2 by one step of the dual simplex algorithm. The resulting graph G is fully defined by the nodes S as well as all connections Γ, i.e. $G = (S, \Gamma)$.

Theorem 2.2 ***(The Connected-graph Theorem)*** *Consider the solution to an mp-LP problem and let θ_1, $\theta_2 \in \Theta_f$ be two arbitrary feasible parameters and $k_1 \in S$ be given such that $\theta_1 \in CR_1$. Then there exists a path $\{k_1, \dots, k_j\}$ in the mp-LP graph $G = (S, \Gamma)$ such that $\theta_2 \in CR_j$.*

Proof: Consider the line segment joining θ_1 and θ_2, i.e.

$$\theta(t) = \theta_1 + t(\theta_2 - \theta_1), \qquad 0 \leq t \leq 1. \tag{2.10}$$

Based on Theorem 2.1, $\theta(t) \in \Theta_f$, as Θ_f is convex. Setting $\theta = \theta(t)$ in the mp-LP problem (2.2) converts the original mp-LP problem into a parametric linear programming (p-LP) problem. The solution of this p-LP problem is given by a series of line segments that are connected as the union constitutes the feasible parameter space Θ_f and is convex based on Theorem 2.1. Based on Eq. (2.6), the limits of each line segment result from the violation of a currently inactive constraint for the parametric solution of that line segment, as all the active constraints are satisfied by definition. Thus, in order to move beyond these limits, the violated constraint needs to be considered as an active constraint, i.e. a step of the dual simplex algorithm needs to be performed. This results in a new active set, and by extension in a sequence of active sets that correspond to the path $\{k_1, \dots, k_j\}$ in the mp-LP graph G. □

In order to visualize the concept described in Theorem 2.2, Figure 2.3 shows a schematic representation of a connected graph for an mp-LP problem.

The proof of Theorem 2.2 is based on the statement that only a single currently inactive constraint limits the solution of the p-LP problem resulting from the substitution of the parametrized line segment joining θ_1 and θ_2. However, in the case of degeneracy or identical constraints, this may not necessarily be the case. While the case of identical constraints can be handled directly, the issue of degeneracy has to be considered in more detail and is discussed in Chapter 2.2.

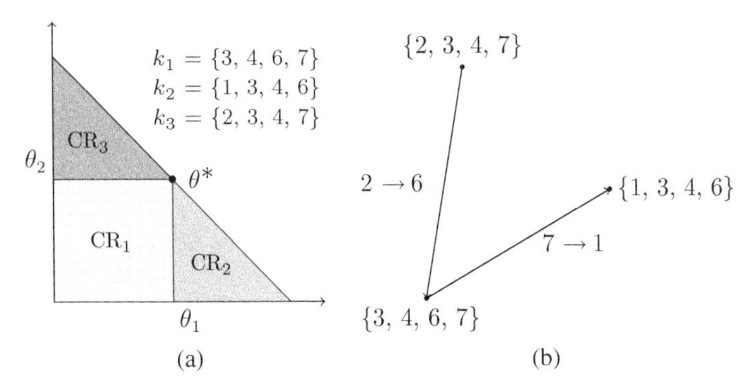

Figure 2.3 A schematic representation of the connected-graph theorem, (a) from a geometrical viewpoint, i.e. considering the a geometric interpretation of the feasible parameter space Θ_f, and (b) from an active set viewpoint, where the dual pivot can be identified in the transition between the active sets associated with the critical regions. Note that although CR_2 and CR_3 have the point θ^* in common, i.e. "k_2 and k_3 are both optimal active sets" (Definition 2.1), it is not possible to pass from k_2 to k_3 in a single step of the dual simplex algorithm. Thus, CR_2 and CR_3 are not connected.

2.2 Degeneracy

The properties described in the previous section are based on the assumption that the active set of the LP problem solved at θ_0 is unique. This uniqueness can only be guaranteed if the solution of the LP problem is non-degenerate. In general, degeneracy refers to the situation where the LP problem under consideration has a specific structure, which does not allow for the unique identification of the active set k.[3] Commonly, the two types of degeneracy encountered are primal and dual degeneracy (see Figure 2.4):

Primal degeneracy: In this case, the vertex of the optimal solution of the LP is overdefined, i.e. there exist multiple sets $k_1 \neq k_2 \neq \cdots \neq k_{\text{tot}}$ such that

$$x_{k_1} = x_{k_2} = \cdots = x_{k_{\text{tot}}}, \tag{2.11}$$

3 This does not consider problems arising from scaling and/or round-off computational errors.

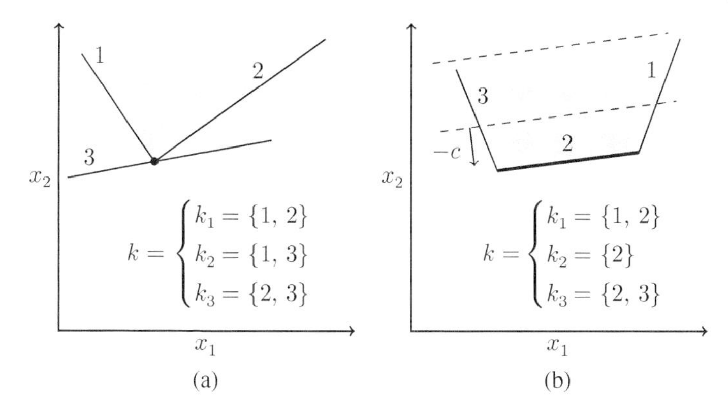

Figure 2.4 Primal and dual degeneracy in linear programming. In (a), primal degeneracy occurs since there are three constraints that are active at the solution, while in (b) dual degeneracy occurs since there is more than one point (x_1, x_2), which features the optimal objective function value.

where, based on Eq. (2.3a):

$$x_k = A_k^{-1}(b_k + F_k\theta_0). \tag{2.12}$$

Dual degeneracy: If there exists more than one point, which attains the optimal objective function value, then the optimal solution x is not unique. Thus, there exist multiple sets $k_1 \neq k_2 \neq \cdots \neq k_{\text{tot}}$ with $x_{k_1} \neq x_{k_2} \neq \cdots \neq x_{k_{\text{tot}}}$ such that

$$z_{k_1} = z_{k_2} = \cdots = z_{k_{\text{tot}}}, \tag{2.13}$$

where $z_k = c^T x_k$. Note that as shown in Figure 2.4, the solution of the problem does not necessarily have to be a vertex, and thus, Eq. (2.12) does not have to hold.

While any solution x_k is a valid solution of the LP problem at θ_0, the key challenge is to identify the effect of primal and dual degeneracy onto the solution of the mp-LP problem.

2.2.1 Primal Degeneracy

Primal degeneracy is caused by the presence of weakly redundant constraints, i.e. constraints that are redundant yet intersect with the feasible parameter space (see Chapter 1.3). In particular, the space where the weakly redundant constraints hold as equality is lower-dimensional with respect to the overall feasible parameter space. Thus, it is clear that if any weakly redundant constraint is chosen as an element of the active set, then the resulting critical region will be lower-dimensional.[4] As a consequence, from all possible combinations of active sets at a given solution, only one active set k_i exists, which results in a full-dimensional critical region, which does not feature any weakly redundant constraints. In the case of Figure 2.4, $k_i = k_1 = \{1, 2\}$. Note that the presence of a lower-dimensional critical region can be detected by calculating the radius of the Chebyshev ball (see Chapter 1.3).

2.2.2 Dual Degeneracy

In general, the effect of primal degeneracy onto the solution of mp-LP problems is manageable, since it can be detected by solving a single LP problem for each candidate active set combination. However, dual degeneracy is much more challenging as the different active sets may result in full-dimensional, but overlapping, critical regions. In particular since the optimal solutions x_k differ, the presence of dual degeneracy might eliminate the continuous nature of the optimizer described in Theorem 2.1.

Remark 2.4 Dual degeneracy results from the non-uniqueness of the optimal solution x_k. This in turn can only occur, since LP problems are not strictly convex, and thus the minimizer is not guaranteed to be unique. Thus, dual degeneracy is not encountered for strictly convex problems such as strictly convex multi-parametric quadratic programming (mp-QP) problems.

However, in order to generate continuous optimizers as well as non-overlapping critical regions, three different approaches have been developed:

---

4 Consider Figure 2.4: if the constraint, which only coincides at the single point with the feasible space is chosen as part of the active set, the corresponding parametric solution from Eq. (2.5) will only be valid in that point, based on Eq. (2.6).

Reformulation to an mp-QP problem [1]: The key idea is to reformulate the mp-LP problem (2.2) into an mp-QP problem (3.2), which yields the same solution at the considered point. Since mp-QP problems do not encounter dual degeneracy due to the inherently unique nature of their optimizers, the continuity of the optimizer can be guaranteed.

Graph/Cluster evaluation [2, 3]: In [2], it was shown that the solution to an mp-LP problem is given by a connected graph, where the nodes are the different active sets and the connections are given by the application of a single step of the dual simplex algorithm. In addition, [3] considers the dual of the mp-LP problem as a parametrized vertex problem and identifies clusters of connected vertices equivalent to the connections in [2]. When dual degeneracy occurs, multiple disconnected graphs/clusters can occur, only some of which may represent the continuous solution of the mp-LP problem across the entire feasible parameter space [3].

Lexicographic perturbation [4]: The problem of dual-degeneracy only arises because of the specific numerical structure of the objective function and the constraints. In order to overcome the degeneracy, the right-hand side of the constraints as well as the objective function are symbolically perturbed in order to obtain a single, continuous optimizer for the solution of the mp-LP problem. Note that the problem is not actually perturbed, but only the result of a proposed perturbation is analyzed enabling the formulation of a continuous optimizer.

2.2.3 Connections Between Degeneracy and Optimality Conditions

Lastly, it is important to highlight that the impact of degeneracy on mp-LP problems goes beyond the derivation of more sophisticated solution strategies. In fact, the presence of primal and dual degeneracy can be directly linked to the optimality conditions required for the calculation of the parametric solution. In the following text, each optimality condition required for the basic sensitivity theorem is revisited with the consideration of the presence of degeneracy.

Second-order sufficient conditions (SOSC): This condition states that the second derivative of the Lagrange function with respect to the optimization variables has to be positive semi-definite. For mp-LP (and convex mp-QP) problems, this condition is naturally satisfied.

Linear Independence Constraint Qualification (LICQ): This condition states that the matrix A_{AS} in Eq. (2.4a) has to have rank n, i.e. there cannot be linearly dependent constraints within the active set. Consider now the case of primal degeneracy, where the number of candidate constraints for the active set $p > n$, i.e. more than n constraints are active at the optimal solution. Clearly, the matrix A_p cannot have full rank, since the maximum rank is n as $x \in \mathbb{R}^n$. As a result, the occurrence of primal degeneracy can be viewed as a LICQ violation at the optimal solution.

Strict Complementary Slackness (SCS): This condition states that there cannot exist a constraint j such that $\lambda_j = 0$ and $A_j x = b_j + F_j \theta_0$. In particular, consider that the Lagrange multiplier λ can be viewed as a "cost" incurred in the objective function when moving along a given constraint. However, dual degeneracy inherently implies that there are multiple points along the same constraints that have the same optimal objective function and thus, this "cost" is equal to 0 (see Figure 2.4 b). Hence, the presence of dual degeneracy is directly linked to the violation of the SCS property, as dual degeneracy implies that there exists at least one constraint j such that $\lambda_j = 0$ and $A_j x = b_j + F_j \theta_0$.

2.3 Critical Region Definition

In linear programming (LP), the term "basic solution" is a result of the use of the simplex algorithm and identifies the solution as a vertex of the feasible space, which is uniquely defined by the indices of the constraints that form the vertex. However, with the emergence of interior-point methods, as well as in the face of degeneracy, it cannot be guaranteed that the solution obtained from an LP solver is a basic solution leading to a full-dimensional critical region. As the classical definition of the critical region is directly tied to the active set (i.e. the indices of the constraints that form the vertex), alternative definitions of critical regions have been considered.

The main theme is thereby to identify an appropriate invariancy set over the parametric space. The three sets typically considered are the following [5]:

Optimal basis invariancy [6]: This invariancy refers to the classical definition of the critical region as a set of active constraints that form a basic solution. The main issue with this approach occurs in the case of degeneracy (see section 2.2), which might lead to lower-dimensional or overlapping regions.

Support set invariancy [7–9]: Given the LP problem formulation

$$\begin{aligned} \underset{x\in\mathbb{R}^n}{\text{Minimize}} \quad & c^T x \\ \text{Subject to} \quad & Ax = b \\ & x \geq 0, \end{aligned} \tag{2.14}$$

the support set is defined as $\sigma(x) = \{i|x_i > 0\}$. The concept of support set invariancy describes the region of the parameter space for which the same support set remains optimal. It can be shown that this eliminates the issue of degeneracy, as the support set is independent of the active constraints.

Optimal partition invariancy [8, 10–12]: The optimal partition is given by the cone, which is spanned from the solution found in the directions of the inactive constraints.

2.4 An Example: Chicago to Topeka

In order to illustrate the concepts developed in this chapter, a classical shipping problem is considered (adapted from [13] and modified for illustrative purposes):

> Given a set of plants $\mathcal{N}$ and a set of markets $\mathcal{M}$ with corresponding supply and demand, and the distances between $\mathcal{N}$ and $\mathcal{M}$, minimize the total transportation cost.

In particular, the transport to Chicago and Topeka is considered, and the problem-specific data is given in Tables 2.1 and 2.2. Note

Table 2.1 The supply and demand of each plant and market in cases.

Supply		Demand	
Seattle	350	Chicago	300
San Diego	600	Topeka	275

Table 2.2 The distances between each plant and market in thousands of miles.

-	Chicago	Topeka
Seattle	1.7	1.8
San Diego	1.8	1.4

that the freight cost is $90 per 1000 miles and case and the freight loading cost is $25 per case.

2.4.1 The Deterministic Solution

Equivalently to the beginning of this chapter, we first consider the uncertainty-free case. Then, the overall transportation cost per case is determined by

$$c_{i,j} = c_{\text{load}} + c_{\text{dist}} \cdot d_{i,j}, \quad i \in \mathcal{N}, \; j \in \mathcal{M} \tag{2.15}$$

where c_{load} is the loading cost, c_{dist} is the distance related cost, and $d_{i,j}$ is the distance between plant and market according to Table 2.1. Thus, the amount of product shipped for all combinations needs to be determined. As there are two plants and two markets, this results in four variables, and a total cost given by:

$$c^T x = \begin{bmatrix} 178 & 187 & 187 & 151 \end{bmatrix} \begin{bmatrix} x_{\text{Se,Ch}} \\ x_{\text{Se,To}} \\ x_{\text{SD,Ch}} \\ x_{\text{SD,To}} \end{bmatrix}, \tag{2.16}$$

where the coefficients are calculated according to Eq. (2.15). The next step is to formulate the constraints. The constraints are that

(i) there cannot be more supply than amount in stock and (ii) the market demands needs to be satisfied. Mathematically, this can be written as

$$\text{Supply limit} \quad \begin{cases} x_{\text{Se,Ch}} + x_{\text{Se,To}} \leq 350 \\ x_{\text{SD,Ch}} + x_{\text{SD,To}} \leq 600 \end{cases} \tag{2.17a}$$

$$\text{Market demand} \quad \begin{cases} x_{\text{Se,Ch}} + x_{\text{SD,Ch}} \geq 300 \\ x_{\text{Se,To}} + x_{\text{SD,To}} \geq 275 \end{cases} \tag{2.17b}$$

Additionally, note that transport can only be positive. This results in the LP problem of the form:

$$\begin{aligned} z^* = \underset{x}{\text{minimize}} \ & c^T x \\ \text{subject to} \ & x_{\text{Se,Ch}} + x_{\text{Se,To}} \leq 350 \\ & x_{\text{SD,Ch}} + x_{\text{SD,To}} \leq 600 \\ & -x_{\text{Se,Ch}} - x_{\text{SD,Ch}} \leq -300 \\ & -x_{\text{Se,To}} - x_{\text{SD,To}} \leq -275 \\ & x = [x_{\text{Se,Ch}}, x_{\text{Se,To}}, x_{\text{SD,Ch}}, x_{\text{SD,To}}]^T \geq 0, \end{aligned} \tag{2.18}$$

the solution of which features the minimal cost of $z^* = \$94\ 925$, and the corresponding transport amounts as

$$x = \begin{bmatrix} 300 \\ 0 \\ 0 \\ 275 \end{bmatrix}. \tag{2.19}$$

2.4.2 Considering Demand Uncertainty

In reality, the data in Table 2.1 is time-varying. Thus, the case of demand uncertainty is considered:

> Given a set of plants $\mathcal{N}$ with a constant supply and a set of markets $\mathcal{M}$ with an uncertain demand bound between 0 and 1000 cases, and the distances between $\mathcal{N}$ and $\mathcal{M}$, minimize the total transportation cost as a function of the demand.

Based on the LP problem (2.18), the following mp-LP problem is formulated:

$$
\begin{aligned}
z(\theta) = \underset{x\in\mathbb{R}^4}{\text{Minimize}} \;\; & c^T x \\
\text{Subject to} \;\; & x_{\text{Se,Ch}} + x_{\text{Se,To}} \leq 350 \\
& x_{\text{SD,Ch}} + x_{\text{SD,To}} \leq 600 \\
& -x_{\text{Se,Ch}} - x_{\text{SD,Ch}} \leq -\theta_1 \\
& -x_{\text{Se,To}} - x_{\text{SD,To}} \leq -\theta_2 \\
& x = [x_{\text{Se,Ch}}, x_{\text{Se,To}}, x_{\text{SD,Ch}}, x_{\text{SD,To}}]^T \geq 0. \\
& \theta = [\theta_1, \theta_2]^T \in \{\theta | 0 \leq \theta_i \leq 1000, i = 1, 2\}
\end{aligned}
\tag{2.20}
$$

Note that the objective function does not change as the distances between the destinations do not change. However on the contrary to problem (2.18), problem (2.20) now features the uncertain demands as parameters $\theta = [\theta_1, \theta_2]^T$ bound between 0 and 1000. The solution of problem (2.20) is reported in Table 2.3, which was obtained based on the solution strategies presented in Chapter 4.

Remark 2.5 The numbering of the constraints is according to their occurrence in problem (2.20), e.g. $x_{\text{Se,Ch}} \geq 0$ is constraint number 5.

2.4.3 Interpretation of the Results

In Table 2.3, it is shown that the solution to problem (2.20) is given by three critical regions. This means that for different demand values, there is a change in which of the constraints in problem (2.20) is active.

The first critical region: For the first critical region, the third and fourth constraints, together with the non-negativity of x, are active. These are the constraints that define the market demand, given from the deterministic values in Eq. (2.17b). Thus, the solution within the first critical region is only concerned with fulfilling the market demand, as the supply limits are not relevant (see second and third critical regions). Consequently, the optimal solution goes along the cheapest transportation

Table 2.3 The solution of problem (2.20).

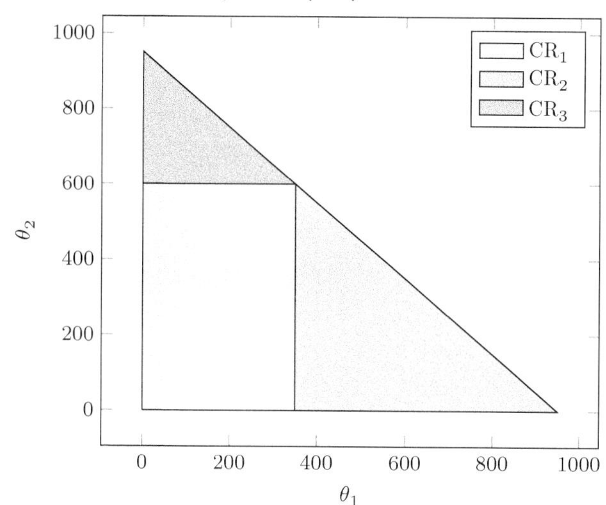

$$\text{CR}_1 \quad \begin{cases} \text{CR} = \{\theta | 0 \leq \theta_1 \leq 350, 0 \leq \theta_2 \leq 600\} \\ k = \{3,\ 4,\ 6,\ 7\} \\ x\,(\theta) = [\theta_1, 0, 0, \theta_2]^T \\ \lambda_k\,(\theta) = [308.31, 261.54, 36, 9]^T \\ z\,(\theta) = 178\theta_1 + 151\theta_2 \end{cases}$$

$$\text{CR}_2 \quad \begin{cases} \text{CR} = \{\theta | \theta_1 \geq 350,\ \theta_2 \geq 0,\ \theta_1 + \theta_2 \leq 950\} \\ k = \{1,\ 3,\ 4,\ 6\} \\ x\,(\theta) = [350, 0, \theta_1 - 350, \theta_2]^T \\ \lambda_k\,(\theta) = [12.73, 323.89, 261.54, 45]^T \\ z\,(\theta) = 187\theta_1 + 151\theta_2 - 3150 \end{cases}$$

$$\text{CR}_3 \quad \begin{cases} \text{CR} = \{\theta | \theta_1 \geq 0,\ \theta_2 \geq 600,\ \theta_1 + \theta_2 \leq 950\} \\ k = \{2,\ 3,\ 4,\ 7\} \\ x\,(\theta) = [\theta_1, \theta_2 - 600, 0, 600]^T \\ \lambda_k\,(\theta) = [50.91, 308.31, 323.89, 45]^T \\ z\,(\theta) = 178\theta_1 + 187\theta_2 - 21600 \end{cases}$$

routes, which are Seattle to Chicago (cost in objective function: 178) and San Diego to Topeka (cost in objective function: 151). The amount that is transported is thereby given by the market demand, i.e. $x_{\text{Se,Ch}} = \theta_1$ and $x_{\text{SD,To}} = \theta_2$.

The second critical region: The second critical region is obtained, when the market demand of Chicago, θ_1, exceeds the supply of Seattle, which is 350. This is apparent in the new active set, which includes the supply constraint of Seattle (the first constraint). Then, in order to fulfill the demand, there needs to be a supply from San Diego, and thus $x_{\text{Se,Ch}} = 350$, $x_{\text{SD,Ch}} = \theta_1 - 350$, while the demand from Topeka can still be fulfilled from San Diego with $x_{\text{SD,To}} = \theta_2$.

The third critical region: Similarly to the second critical region, the third critical region results when San Diego is unable to meet all the demands from Topeka, as the supply limit of 600 is reached. Then, the supply constraint from San Diego becomes active (the second constraint), and in order to fulfill the demand, material will be transported from Seattle to Topeka, and thus $x_{\text{SD,To}} = 600$, $x_{\text{Se,To}} = \theta_2 - 600$, while the demand from Chicago is met from Seattle with $x_{\text{Se,Ch}} = \theta_1$.

Infeasible region: As soon as the sum of the demand, $\theta_1 + \theta_2$, is greater than the available supply, $600 + 350 = 950$, there is no possibility to meet all the demands with the supply given. Thus, there is no feasible solution for $\theta_1 + \theta_2 > 950$.

2.5 Literature Review

The idea to consider the variation of a parameter in an LP problem was reportedly first put forth in the unpublished master thesis by William Orchard-Hays in 1952, as reported in [14], where the variation of the right-hand side of the constraints was considered:

$$\sum_{j=1}^{n} a_{ij}x_j = b_i + \theta b'_i, \quad \forall i = 1, \ldots, m. \tag{2.21}$$

Subsequently, several researchers proposed algorithms for the treatment of the single parameter case [15–20], but it was not until 1972 when the seminal paper by Gal and Nedoma described for the first time a rigorous strategy for the general solution of mp-LP problems of type (2.2) based on the connected-graph theorem [2]. All these developments were captured in the excellent textbook by Gal and Davis from 1979 [21], for which a second edition appeared in 1995 [22].

The algorithm from [2] remained the state-of-the-art for over 25 years (see, e.g. [23, 24]), until geometrical algorithms for general mp-LP problems were developed starting in 2000,[5] with the application of multi-parametric programming to model-predictive control [26, 27]. Although the initial focus was put on mp-QP problems, quickly publications concerning combination of model predictive control and mp-LP problems were put forth [28–31], many of which were captured in this excellent review [32]. This new string of developments resulted in a deeper interest in the theoretical properties of mp-LP problems, specifically in the case of degeneracy (see section 2.2), as well as the question of the presence of parameters in the left-hand side of the constraints, i.e.

$$A(\theta)x \leq b + F\theta \tag{2.22a}$$

$$A(\theta) := A_N + \sum_{i=1}^{q} \theta_i A_i, \tag{2.22b}$$

where algorithms based on McCormick relaxations [33, 34], as well as exact algorithms for the single parameter case [35] have been presented.[6]

References

1 Spjøtvold, J., Tøndel, P., and Johansen, T.A. (2005) A method for obtaining continuous solutions to multiparametric linear programs, in World Congress, IFAC, Elsevier, IFAC proceedings volumes, p. 902, doi: 0703-6-CZ-1902.00903.

5 The geometrical algorithms presented up to that point were limited to at most two parameters [2, 25].
6 In his book, Gal also considered the case of left-hand side uncertainty, however limited to a single parameter and a single row or column [22].

2 Gal, T. and Nedoma, J. (1972) Multiparametric linear programming. *Management Science*, 18 (7), 406–422, doi: 10.1287/mnsc.18.7.406.

3 Olaru, S. and Dumur, D. (2006) On the continuity and complexity of control laws based on multiparametric linear programs, in *45th IEEE Conference on Decision and Control, 2006*, pp. 5465–5470, doi: 10.1109/CDC.2006.377330.

4 Jones, C.N., Kerrigan, E.C., and Maciejowski, J.M. (2007) Lexicographic perturbation for multiparametric linear programming with applications to control. *Automatica*, 43 (10), 1808–1816, doi: 10.1016/j.automatica.2007.03.008. URL http://www.sciencedirect.com/science/article/pii/S0005109807002002.

5 Hladík, M. (2010) Multiparametric linear programming: support set and optimal partition invariancy. *European Journal of Operational Research*, 202 (1), 25–31, doi: 10.1016/j.ejor.2009.04.019. URL http://www.sciencedirect.com/science/article/pii/S0377221709002926.

6 Gal, T. and Greenberg, H.J. (1997) *Advances in sensitivity analysis and parametric programming*, vol. 6, Springer US, Boston, MA, doi: 10.1007/978-1-4615-6103-3.

7 Hadigheh, A.G. and Terlaky, T. (2006) Generalized support set invariancy sensitivity analysis in linear optimization. *Journal of Industrial and Management Optimization*, 2 (1), 1–18.

8 Hadigheh, A.G. and Terlaky, T. (2006) Sensitivity analysis in linear optimization: invariant support set intervals. *European Journal of Operational Research*, 169 (3), 1158–1175, doi: 10.1016/j.ejor.2004.09.058. URL http://www.sciencedirect.com/science/article/pii/S0377221705002808.

9 Hadigheh, A.G., Mirnia, K., and Terlaky, T. (2007) Active constraint set invariancy sensitivity analysis in linear optimization. *Journal of Optimization Theory and Applications*, 133 (3), 303–315, doi: 10.1007/s10957-007-9201-5. URL http://dx.doi.org/10.1007/s10957-007-9201-5.

10 Greenberg, H.J. (1994) The use of the optimal partition in a linear programming solution for postoptimal analysis. *Operations Research Letters*, 15 (4), 179–185, doi: 10.1016/0167-6377(94)90075-2. URL http://www.sciencedirect.com/science/article/pii/0167637794900752.

11 Berkelaar, A.B., Roos, K., and Terlaky, T. (1997) The optimal set and optimal partition approach to linear and quadratic programming, in (eds T. Gal and H.J. Greenberg) *Advances in sensitivity analysis and parametic programming*, Springer US, Boston, MA, pp. 159–202, doi: 10.1007/978-1-4615-6103-3_6. URL http://dx.doi.org/10.1007/978-1-4615-6103-3_6.

12 Greenberg, H.J. (2000) Simultaneous primal-dual right-hand-side sensitivity analysis from a strictly complementary solution of a linear program. *SIAM Journal on Optimization*, 10 (2), 427–442, doi: 10.1137/S1052623496310333. URL http://dx.doi.org/10.1137/S1052623496310333.

13 Dantzig, G.B. (1963) *Linear programming and extensions*, Princeton University Press, Princeton, NJ.

14 Gal, T. (1985) The historical development of parametric programming, in *Parametric optimization and approximation, International series of numerical mathematics / Internationale Schriftenreihe zur Numerischen Mathematik / Série internationale d'Analyse numérique*, vol. 72 (eds B. Brosowski and F. Deutsch), Birkhäuser Verlag, Basel, pp. 148–165, doi: 10.1007/978-3-0348-6253-0_10. URL http://dx.doi.org/10.1007/978-3-0348-6253-0_10.

15 Manne, A.S. (1953) *Notes on parametric linear programming*, RAND Corporation, pp. P–468. URL http://www.rand.org/pubs/papers/P468.

16 Gass, S. and Saaty, T. (1955) The computational algorithm for the parametric objective function. *Naval Research Logistics Quarterly*, 2 (1–2), 39–45, doi: 10.1002/nav.3800020106. URL http://dx.doi.org/10.1002/nav.3800020106.

17 Orchard-Hays, W. (1955) *The RAND code for the simplex method (SX4): (For the IBM 701 electronic computer)*, Rand Corporation, Santa Monica, CA.

18 Saaty, T.L. (1959) Coefficient perturbation of a constrained extremum. *Operations Research*, 7 (3), 294–302, doi: 10.1287/opre.7.3.294. URL http://dx.doi.org/10.1287/opre.7.3.294.

19 Simons, E. (1962) A note on parametric linear programming. *Management Science*, 8 (3), 355–358, doi: 10.1287/mnsc.8.3.355. URL http://dx.doi.org/10.1287/mnsc.8.3.355.

20 Karabegov, V.K. (1963) A parametric problem in linear programming. *USSR Computational Mathematics and Mathematical Physics*, 3 (3), 725–741, doi: 10.1016/0041-5553(63)90297-0. URL http://www.sciencedirect.com/science/article/pii/0041555363902970.

21 Gal, T. and Davis, G.V. (1978, cop. 1979) *Postoptimal analyses, parametric programming and related topics*, McGraw-Hill, London.

22 Gál, T. (1995) *Postoptimal analyses, parametric programming, and related topics: degeneracy, multicriteria decision making, redundancy*, W. de Gruyter, Berlin and New York, 2nd edn.

23 Acevedo, J. and Pistikopoulos, E.N. (1997) A multiparametric programming approach for linear process engineering problems under uncertainty. *Industrial and Engineering Chemistry Research*, 36 (3), 717–728, doi: 10.1021/ie960451l.

24 Dua, V. and Pistikopoulos, E.N. (1999) Algorithms for the solution of multiparametric mixed-integer nonlinear optimization problems. *Industrial and Engineering Chemistry Research*, 38 (10), 3976–3987, doi: 10.1021/ie980792u.

25 Dinkelbach, W. (1969) *Sensitivitätsanalysen und parametrische Programmierung, Ökonometrie und Unternehmensforschung / Econometrics and Operations Research*, vol. 12, Springer-Verlag, Berlin, Heidelberg.

26 Bemporad, A., Morari, M., Dua, V., and Pistikopoulos, E.N. (2000) The explicit solution of model predictive control via multiparametric quadratic programming. *Proceedings of the American Control Conference*, vol. 2, pp. 872–876, doi: 10.1109/ACC.2000.876624.

27 Bemporad, A., Morari, M., Dua, V., and Pistikopoulos, E.N. (2002) The explicit linear quadratic regulator for constrained systems. *Automatica*, 38 (1), 3–20, doi: 10.1016/S0005-1098(01)00174-1. URL http://www.sciencedirect.com/science/article/pii/S0005109801001741.

28 Bemporad, A., Borrelli, F., and Morari, M. (2002) Model predictive control based on linear programming - the explicit solution. *IEEE Transactions on Automatic Control*, 47 (12), 1974–1985, doi: 10.1109/TAC.2002.805688.

29 Bemporad, A., Borrelli, F., and Morari, M. (2000) The explicit solution of constrained LP-based receding horizon control, in *Proceedings of the 39th IEEE Conference on Decision and Control, 2000*, vol. 1, pp. 632–637, doi: 10.1109/CDC.2000.912837.

30 Borrelli, F., Bemporad, A., and Morari, M. (2003) Geometric algorithm for multiparametric linear programming. *Journal of Optimization Theory and Applications*, 118 (3), 515–540, doi: 10.1023/B:JOTA.0000004869.66331.5c. URL URL http://dx.doi.org/10.1023/B%3AJOTA.0000004869.66331.5c.

31 Morari, M., Jones, C.N., Zeilinger, M.N., and Baric, M. (2008) Multiparametric linear programming for control, in *CCC 2008. 27th Chinese Control Conference, 2008*, pp. 2–4, doi: 10.1109/CHICC.2008.4604876.

32 Jones, C.N., Barić, M., and Morari, M. (2007) Multiparametric linear programming with applications to control. *European Journal of Control*, 13 (2–3), 152–170, doi: 10.3166/ejc.13.152-170. URL http://www.sciencedirect.com/science/article/pii/S0947358007708178.

33 Wittmann-Hohlbein, M. and Pistikopoulos, E.N. (2012) A two-stage method for the approximate solution of general multiparametric mixed-integer linear programming problems. *Industrial and Engineering Chemistry Research*, 51 (23), 8095–8107, doi: 10.1021/ie201408p.

34 Wittmann-Hohlbein, M. and Pistikopoulos, E.N. (2013) On the global solution of multi-parametric mixed integer linear programming problems. *Journal of Global Optimization*, 57 (1), 51–73, doi: 10.1007/s10898-012-9895-2. URL http://dx.doi.org/10.1007/s10898-012-9895-2.

35 Khalilpour, R. and Karimi, I.A. (2014) Parametric optimization with uncertainty on the left hand side of linear programs. *Computers and Chemical Engineering*, 60, 31–40, doi: 10.1016/j.compchemeng.2013.08.005. URL http://www.sciencedirect.com/science/article/pii/S0098135413002421.

3

Multi-Parametric Quadratic Programming

Consider the following quadratic programming (QP) problem:

$$\begin{aligned} z = \underset{x}{\text{Minimize}} \quad & \left(\frac{1}{2}Qx + c\right)^T x \\ \text{Subject to} \quad & Ax \leq b \\ & A_{\text{eq}}x = b_{\text{eq}} \\ & x \in \mathbb{R}^n, \end{aligned} \tag{3.1}$$

where $Q \in \mathbb{R}^{n \times n}$ is symmetric positive definite, $c \in \mathbb{R}^n$, $A \in \mathbb{R}^{m \times n}$, $b \in \mathbb{R}^m$, $A_{\text{eq}} \in \mathbb{R}^{s \times n}$, $b_{\text{eq}} \in \mathbb{R}^s$, $Ax \leq b$ is a compact polytope, and A_{eq} is assumed to have full rank.[1] This type of problems is encountered frequently in areas such as optimal control, scheduling, planning, and resourcing. This is mainly due to the fact that many common metrics such as the minimization of the sum of squares result in a convex quadratic function. However, equivalently to the LP problem (2.1), the QP problem (3.1) assumes that all information is deterministically known, which might not necessarily be the case. Thus, in order to account for this uncertainty, a multi-parametric quadratic programming (mp-QP) problem can be

1 If A_{eq} does not have full rank, it is always possible to find an equivalent matrix A'_{eq} with a reduced number of rows which has full rank.

Multi-parametric Optimization and Control, First Edition.
Efstratios N. Pistikopoulos, Nikolaos A. Diangelakis, and Richard Oberdieck.

formulated:

$$
\begin{aligned}
z(\theta) = \underset{x}{\text{Minimize}} \quad & \left(\frac{1}{2}Qx + H\theta + c\right)^T x \\
\text{Subject to} \quad & Ax \leq b + F\theta \\
& A_{\text{eq}}x = b_{\text{eq}} + F_{\text{eq}}\theta \\
& \theta \in \Theta := \{\theta \in \mathbb{R}^q | \text{CR}_A\theta \leq \text{CR}_b\} \\
& x \in \mathbb{R}^n,
\end{aligned}
\tag{3.2}
$$

where $H \in \mathbb{R}^{n\times q}$, $F \in \mathbb{R}^{m\times q}$, $F_{\text{eq}} \in \mathbb{R}^{s\times q}$, $\text{CR}_A \in \mathbb{R}^{r\times q}$, $\text{CR}_b \in \mathbb{R}^r$, and Θ are a compact polytope. Equivalent to Chapter 2, the difference between problems (3.1) and (3.2) lies in the presence of the uncertain parameter vector θ. As a result, the solution x, the Lagrange multipliers λ and μ, and the optimal objective function z of problem (3.2) are obtained as a function of θ, i.e. $x(\theta)$, $\lambda(\theta)$, $\mu(\theta)$, and $z(\theta)$, respectively. A schematic representation of the difference between QP and mp-QP problems is shown in Figure 3.1.

Remark 3.1 Note that it is possible to add a scalar d to the objective function of a QP problem without influencing the

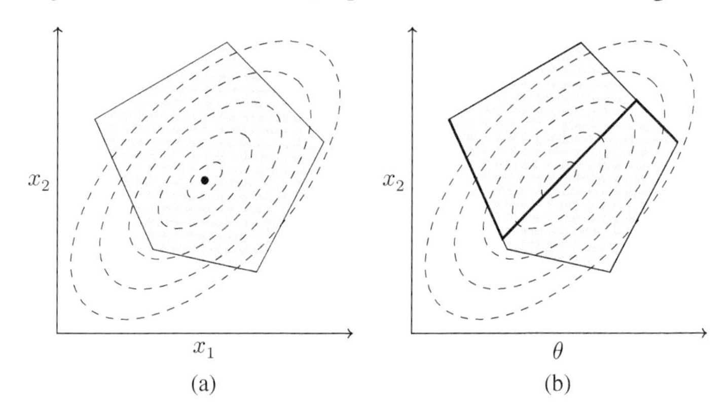

Figure 3.1 The difference between the solution of a QP and an mp-QP problem (black dot and line, respectively), where the mp-QP problem is obtained by treating x_1 as a parameter. In (a), the solution is a single point in one of the vertices of the feasible space, while in (b) the solution is a function of the parameter θ. The parametric solution is given by the unconstrained solution, until a constraint is violated. At that point, it is necessary to explicitly consider the constraint, which yields the shape of the parametric solution.

optimal solution. Similarly, it is possible to add an arbitrary scaling function $f(\theta)$ to an mp-QP problem without influencing the optimal solution.

3.1 Calculation of the Parametric Solution

Remark 3.2 Due to the similarities between mp-LP and mp-QP problems, the different solution strategies available are discussed in detail in Chapter 4.

3.1.1 Solution *via* the Basic Sensitivity Theorem

Let θ be fixed to a nominal parameter value, $\theta = \theta_0$. Then, the solution of the resulting QP problem of the form (3.1) is given by the solution x_0 as well as the values of the Lagrange multipliers λ_0. Based on these, the indices of the active constraints at the solution (also called the active set) k are identified as

$$A_k x_0 = b_k + F_k \theta_0 \tag{3.3a}$$

$$\lambda_{0,k} > 0. \tag{3.3b}$$

Remark 3.3 In the case where the set k from Eq. (3.3) is not unique, the solution is said to be degenerate. The impact of degeneracy on the parametric solution is discussed in Section 2.2.

Based on Chapter 1, this results in the parametric solution:

$$\begin{bmatrix} x(\theta) \\ \lambda_k(\theta) \\ \mu(\theta) \end{bmatrix} = \begin{bmatrix} x_0 \\ \lambda_{0,k} \\ \mu_0 \end{bmatrix} + (M_0)^{-1} N_0 (\theta - \theta_0), \tag{3.4}$$

where

$$M_0 = \begin{bmatrix} Q & A_k^T & A_{eq}^T \\ \lambda A_k & A_k x_0 - b_k - F_k \theta_0 & 0 \\ \mu A_{eq} & 0 & A_{eq} x_0 - b_{eq} + F_{eq} \theta_0 \end{bmatrix} \tag{3.5a}$$

$$N_0 = \begin{bmatrix} H & -\lambda F_k & -\mu F_{eq} \end{bmatrix}^T \tag{3.5b}$$

3.1.2 Solution *via* the Parametric Solution of the KKT Conditions

Alternatively to Eq. (3.4), it is also possible to solve the Karush–Kuhn–Tucker (KKT) conditions in a parametric fashion: let k be the indices of the active constraints according to Eq. (3.3). First, consider the explicit form of the derivative of the Lagrange function $\mathcal{L}(x, \lambda, \mu, \theta)$, i.e.

$$\begin{aligned}\nabla_x \mathcal{L}(x, \lambda, \mu, \theta) &= \nabla_x((Qx + H\theta + c)^T x) \\ &\quad + \nabla_x \left(\sum_{i \in k} \lambda_i (A_i x - b_i - F_i \theta) \right) \qquad (3.6a) \\ &\quad + \nabla_x \left(\sum_{j=1}^{s} \mu_j (A_{\text{eq},j} x - b_{\text{eq},j} - F_{\text{eq},j} \theta) \right) \qquad (3.6b) \\ &= Qx + H\theta + c + A_k^T \lambda_k + A_{\text{eq}}^T \mu. \qquad (3.6c)\end{aligned}$$

Then, the corresponding KKT conditions are given as

$$Qx + H\theta + c + A_k^T \lambda_k + A_{\text{eq}}^T \mu = 0 \qquad (3.7a)$$

$$A_k x - b_k - F_k \theta = 0 \qquad (3.7b)$$

$$A_{\text{eq}} x - b_{\text{eq}} - F_{\text{eq}} \theta = 0, \qquad (3.7c)$$

and it is possible to reformulate Eq. (3.7a) such that

$$x(\theta) = -Q^{-1} \left(H\theta + c + \begin{bmatrix} A_k^T & A_{\text{eq}}^T \end{bmatrix} \begin{bmatrix} \lambda_k \\ \mu \end{bmatrix} \right). \qquad (3.8)$$

Note that Q is invertible since it is positive definite. For convenience, let the active set matrices be defined as

$$A_{\text{AS}} = \begin{bmatrix} A_k^T & A_{\text{eq}}^T \end{bmatrix}^T \qquad (3.9a)$$

$$b_{\text{AS}} = \begin{bmatrix} b_k^T & b_{\text{eq}}^T \end{bmatrix}^T \qquad (3.9b)$$

$$F_{\text{AS}} = \begin{bmatrix} F_k^T & F_{\text{eq}}^T \end{bmatrix}^T. \qquad (3.9c)$$

Then, the substitution of Eq. (3.8) into Eq. (3.7b) results in

$$\begin{aligned}&- A_{\text{AS}} Q^{-1} \left(H^T\theta + c + A_{\text{AS}}^T \begin{bmatrix} \lambda_k \\ \mu \end{bmatrix} \right) - b_{\text{AS}} - F_{\text{AS}} \theta = 0 \\ &\Rightarrow \begin{bmatrix} \lambda_k(\theta) \\ \mu(\theta) \end{bmatrix} = -(A_{\text{AS}} Q^{-1} A_{\text{AS}}^T)^{-1} (b_{\text{AS}} + F_{\text{AS}} \theta \\ &\qquad\qquad + A_{\text{AS}} Q^{-1} (H\theta + c)), \qquad (3.10)\end{aligned}$$

which can be substituted into Eq. (3.8) to obtain the full parametric solution.

3.2 Solution Properties

3.2.1 Local Properties

Based on Eq. (3.4) the following statements about the solution of mp-QP problems can be made:

- The optimization variable $x(\theta)$ is an affine function of the parameter θ.
- The Lagrange multipliers $\lambda_k(\theta)$ and $\mu(\theta)$ of the active constraints are affine functions of the parameter θ.
- In order for Eq. (3.4) to remain the optimal solution around θ_0, it needs to be feasible and optimal, i.e.

$$Ax(\theta) \leq b + F\theta \qquad \text{(Feasibility of } x(\theta)\text{)} \tag{3.11a}$$

$$\lambda_k(\theta) \geq 0 \qquad \text{(Optimality of } x(\theta)\text{)} \tag{3.11b}$$

$$\text{CR}_A\theta \leq \text{CR}_b \qquad \text{(Feasibility of } \theta\text{)} \tag{3.11c}$$

Remark 3.4 Note that due to the fact the equality constraints need to hold for all θ, it is not necessary to include them in the critical region definition.

Thus, around optimal solution of problem (3.2) around θ_0 is given by Eq. (3.4) and is valid in the compact polytope described by Eq. (3.10), which is referred to as *critical region* CR:

$$\text{CR} = \left\{ \theta \in \mathbb{R}^q \;\middle|\; \begin{array}{l} Ax(\theta) \leq b + F\theta \\ \lambda_k(\theta) \geq 0 \\ \text{CR}_A\theta \leq \text{CR}_b \end{array} \right\}. \tag{3.12}$$

Note that Lemma 1 also holds for mp-QP problems, i.e. every critical region is uniquely defined by its active set.

3.2.2 Global Properties

Similarly to the mp-LP case, these solution properties hold for any feasible point θ_0 and thus the following theorem can be formulated:

Theorem 3.1 ***(The Solution of mp-QP Problems)*** *Consider the mp-QP problem (2.2). Then the set of feasible parameters $\Theta_f \subseteq \Theta$ is convex, the optimizer $x(\theta) : \Theta_f \mapsto \mathbb{R}^n$ is continuous and piecewise affine, and the optimal objective function $z(\theta) : \Theta_f \mapsto \mathbb{R}$ is continuous and piecewise quadratic.*

Proof: See Theorem 2.1. □

The concept of the partitioning of the parameter space is schematically represented in Figure 3.2. However, mp-QP problems differ in the following aspects from mp-LP problems (see Figure 3.3):

Convexity: It is important to note that problem (3.2) is only required to be convex in the optimization variable x, but not in the parameter θ. As a result, it cannot be guaranteed that

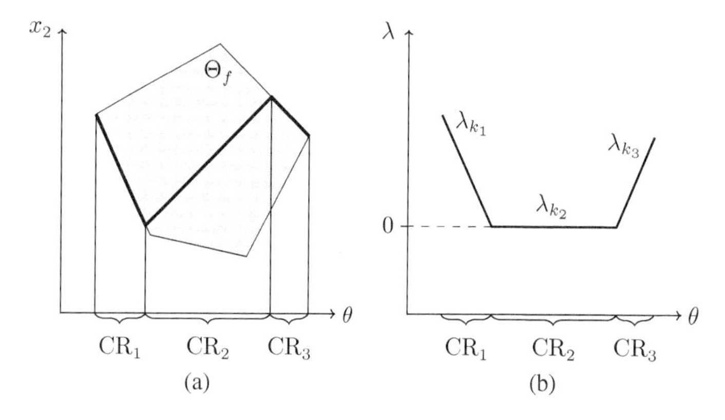

Figure 3.2 A schematic representation of the solution of the mp-QP problem from Figure 3.1. In (a), the partitioning of the convex, feasible parameter space Θ_f, as well as the piecewise affine nature of the optimal solution $x(\theta)$ is shown. In (b), the Lagrange multipliers λ_k as a function of θ are shown. Note that $\lambda_{k_2} = 0$ since the parametric solution features no active constraints, as it is the unconstrained solution.

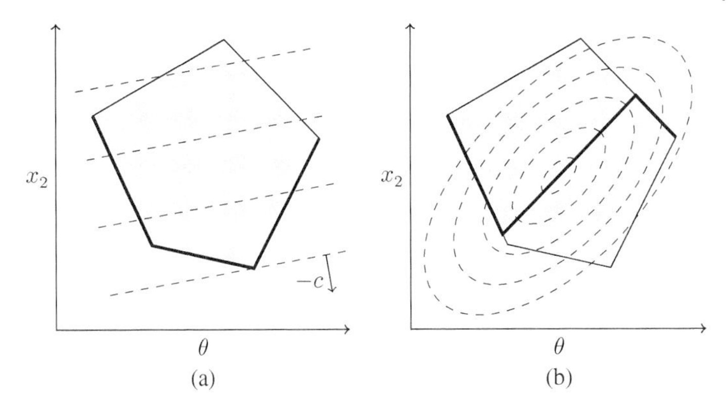

Figure 3.3 A schematic representation of some of the differences between (a) mp-LP problems and (b) mp-QP problems.

the optimal objective function $z(\theta)$ is convex for all possible cross-terms H. However, as seen in Eq. (3.7) this potential non-convexity does not influence the optimality conditions.

Combinatorial complexity: On the contrary to the solution to an LP problem, the solution to a QP problem does not have to lie on a vertex. In fact, the number of active constraints can vary between 0 and n and thus, the maximum number of critical region is bounded from above by $\sum_{i=s}^{n} \binom{m}{i}$.

Dual degeneracy: The mp-QP problem (3.2) is strictly convex if and only if Q is positive definite. A strictly convex mp-QP implies that the minimizer for any fixed $\theta = \theta_0$ is unique and thus, there cannot be any dual degeneracy. This immediately also leads to the conclusion that the strict complementary slackness (SCS) condition is always fulfilled for strictly convex mp-QP problems, since there cannot be a case where $\lambda_k = 0$ and $A_k x = b_k + F_k \theta_0$ due to the uniqueness of the minimizer (see Section 2.2 for a more detailed treatment of degeneracy). Thus, since also the second-order sufficient conditions (SOSC) are always fulfilled, the only condition that needs to be verified for strictly convex mp-QP problems is the linear independence constraint qualification (LICQ), which is discussed in detail in Section 3.2.3.

3.2.3 Structural Analysis of the Parametric Solution

Similarly to the solution of the mp-LP problem (2.2), the solution of the mp-QP problem (3.2) features several structural properties, which provide insights into the relations between different critical regions.

Theorem 3.2 *(Active Set of Adjacent Region) Consider an active set $k = \{i_1, i_2, \dots, i_k\}$ and its corresponding critical region CR_0 in minimal representation, i.e. with all redundant constraints removed. Additionally, let CR_i be a full-dimensional neighboring critical region to CR_0 and assume that the linear independent constraint qualification holds on their common facet $F = \mathrm{CR}_0 \cap H$, where H is the separating hyperplane, and let $\lambda(\theta)$ be the Lagrange multiplier. Moreover, assume that there are no constraints, which are weakly active at the optimizer $x(\theta)$ for all $\theta \in \mathrm{CR}_0$. Then*

Type I: *If H is given by $A_{i_{k+1}} x(\theta) = b_{i_{k+1}} + F_{i_{k+1}} \theta$, then the optimal active set in CR_i is $\{i_1, \dots, i_k, i_{k+1}\}$.*

Type II: *If H is given by $\lambda_{i_k}(\theta) = 0$, then the optimal active set in CR_i is $\{i_1, \dots, i_{k-1}\}$.*

Proof: The first thing to prove is Type I. In order for some constraint $i_j \in k$ not to be in the active set of CR_i, $\lambda_{i_j}(\theta) = 0$ for all $\theta \in \mathcal{F}$. However, since there are no weakly active constraints for all $x \in \mathrm{CR}_0$, this would mean that the constraint i_j becomes inactive at $\mathcal{F}$. This contradicts the assumption of minimality since $\lambda_{i_j}(\theta) \geq 0$ and $A_{i_{|k|+1}} x(\theta) = b_{i_{|k|+1}} + F_{i_{|k|+1}} \theta$ would be coincident. Conversely, k cannot be the active set of CR_i due to the uniqueness of the active set for each critical region. Thus, the active set of CR_i needs to be a superset of k. Assume now that another constraint $i_{|k|+2}$ is active in CR_i. This implies that $A_{i_{|k|+2}} x(\theta) = b_{i_{|k|+2}} + F_{i_{|k|+2}} \theta$ for all $\theta \in \mathrm{CR}_i$ and in particular for all $\theta \in \mathcal{F}$. However, this means that $A_{i_{|k|+1}} x(\theta) = b_{i_{|k|+1}} + F_{i_{|k|+1}} \theta$ and $A_{i_{|k|+2}} x(\theta) = b_{i_{|k|+2}} + F_{i_{|k|+2}} \theta$ would coincide, which contradicts the assumption of minimality. Thus, only $\{i_1, i_2, \dots, i_{|k|}, i_{|k|+1}\}$ can be the active set of CR_i. The proof for Type II is similar. □

Lemma 3.1 *The facet-to-facet property for facet i of* CR_0 *in minimal representation holds if LICQ and SCS are fulfilled for all* $\theta \in int(\mathcal{F})$, *where int denotes the interior of a set,* $\mathcal{F} = \text{CR}_0 \cap \mathcal{H}$ *and* $\mathcal{H}$ *is the separating hyperplane of the ith facet.*

Proof: Equivalently to the proof of Theorem 3.2, the assumption of minimal representation of CR_0 requires that Theorem 3.2 holds for all points in the interior of the considered facet. □

Lemma 3.2 *Let the conditions of Theorem 3.2 hold. Then the Lagrange multipliers* λ *are a continuous, piecewise affine function of the parameter* θ.

Proof: The piecewise affine nature of the Lagrange function follows directly from Eq. (3.10). In order to prove the continuity under the conditions of Theorem 3.2, consider the ith constraint, two adjacent critical regions CR_1 and CR_2, and assume that the lemma is incorrect, i.e. there exists a point θ^* on the interior of the facet such that λ_i is not continuous. Then, this would imply the existence of a non-unique solution of the dual problem, which is possible if and only if the problem is degenerate. This however contradicts the conditions of Theorem 3.2 and completes the proof. □

Based on these developments, the connected-graph theorem for mp-LP problems, Theorem 2.2, is extended to the case of mp-QP problems.

Definition 3.1 (mp-QP Graph).Let each optimal active set k of an mp-QP problem be a node in $\mathcal{S}$. Then the nodes k_1 and k_2 are connected if (i) there exists $\theta^* \in \Theta_f$ such that k_1 and k_2 are both optimal active sets and (ii) the conditions of Theorem 3.2 are fulfilled on the facet or it is possible to pass from k_1 to k_2 by one step of the dual simplex algorithm. The resulting graph G is fully defined by the nodes $\mathcal{S}$ as well as all connections Γ, i.e. $G = (\mathcal{S}, \Gamma)$.

Theorem 3.3 ***(Connected Graph for mp-QP Problems)*** *Consider the solution to an mp-QP problem and let* $\theta_1, \theta_2 \in \Theta_f$ *be* *two arbitrary feasible parameters and* $k_1 \in \mathcal{S}$ *be given such that* $\theta_1 \in \mathrm{CR}_1$. *Then there exists a path* $\{k_1, \dots, k_j\}$ *in the mp-QP graph* $G = (\mathcal{S}, \Gamma)$ *such that* $\theta_2 \in \mathrm{CR}_j$.

Proof: If the conditions of Theorem 3.2 are fulfilled, then it is clear that a connected graph results. As Theorem 3.2 does not hold if either LICQ does not hold or weakly active constraints are present, we have to prove that a step of the dual simplex algorithm is enough to identify all candidates of the adjacent region. First, as strictly convex mp-QP problems do not feature weakly active constraints, only possible violations of LICQ need to be considered. The LICQ violation can only occur in Type I constraints of Theorem 3.2, since a constraint is added and thus the linearly independent nature of the candidate active set might change. In the case where the cardinality of the original active set is n, Theorem 3 holds directly. If the cardinality is less than n but LICQ is violated this means that an equivalent, lower-dimensional problem can be formulated where Theorem 3 holds as well, resulting in a connected graph. □

This theorem has the following implications:

- If the conditions of Theorem 3.2 are fulfilled, only one candidate active set is generated.
- Let k be an active set with cardinality p. If LICQ is violated on the border of the corresponding critical region, the number of candidates is given as $\binom{p}{p-1}$.

In order to visualize the concept described in Theorem 3.3, Figure 3.4 shows a schematic representation of a connected graph for an mp-QP problem.

The proof of Theorem 3.3 is based on the statement that only a single currently inactive constraint limits the solution of the mp-QP problem. However, in the case of degeneracy or identical constraints, this may not necessarily be the case. While the case of identical constraints can be handled directly, the issue of degeneracy has to be considered in more detail and is discussed in Section 2.2.

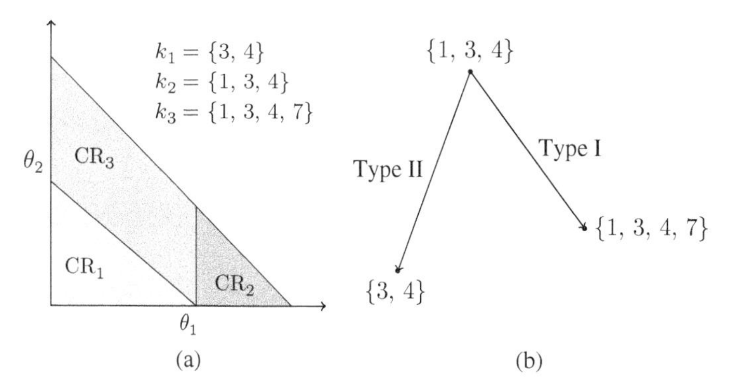

Figure 3.4 A schematic representation of the connected-graph theorem, (a) from a geometrical viewpoint, i.e. considering the a geometric interpretation of the feasible parameter space Θ_f, and (b) from an active set viewpoint, where the type of adjacency according to Theorem 3.2 can be identified in the transition between the active sets associated with the critical regions.

3.3 Chicago to Topeka with Quadratic Distance Cost

In Chapter 2, a shipping problem was introduced where the demand uncertainty was considered by formulating an mp-LP problem. Thereby, it was assumed that the price per distance is linear. In this chapter, the case of a quadratic distance cost of \$90 per 1000 miles and case squared is considered. Thus, the overall cost function $J(x)$ is given as

$$J(x) = x^T \operatorname{diag}(c_{\mathrm{dist}} \cdot d_{ij})x + c_{\mathrm{load}}^T x$$

$$= x^T \begin{bmatrix} 153 & 0 & 0 & 0 \\ 0 & 162 & 0 & 0 \\ 0 & 0 & 162 & 0 \\ 0 & 0 & 0 & 126 \end{bmatrix} x + \begin{bmatrix} 25 & 25 & 25 & 25 \end{bmatrix} x, \tag{3.13}$$

where $x = [x_{\mathrm{Se,Ch}}, x_{\mathrm{Se,To}}, x_{\mathrm{SD,Ch}}, x_{\mathrm{SD,To}}]^T$ and $\operatorname{diag}(a)$ denotes a diagonal matrix where the weights are specified by the vector a.

The corresponding shipping problem from Chapter 2 is transformed into the following mp-QP problem:

$$
\begin{aligned}
z(\theta) = \underset{x\in\mathbb{R}^4}{\text{minimize}}\ & J(x) \\
\text{subject to}\ & x_{\text{Se,Ch}} + x_{\text{Se,To}} \leq 350 \\
& x_{\text{SD,Ch}} + x_{\text{SD,To}} \leq 600 \\
& -x_{\text{Se,Ch}} - x_{\text{SD,Ch}} \leq -\theta_1 \\
& -x_{\text{Se,To}} - x_{\text{SD,To}} \leq -\theta_2 \\
& x = [x_{\text{Se,Ch}}, x_{\text{Se,To}}, x_{\text{SD,Ch}}, x_{\text{SD,To}}]^T \geq 0 \\
& \theta = [\theta_1, \theta_2]^T \in \{\theta \in \mathbb{R}^2 | 0 \leq \theta_i \leq 1000, i = 1,2\}.
\end{aligned}
\tag{3.14}
$$

Note that the constraints remain unchanged, which implies that the feasible space will also not change. The solution of problem (3.14) is reported in Table 3.1 and a graphical representation of the geometrical view as well as the active set view is shown in Figure 3.5. Note that the solution was obtained based on the solution strategies presented in Chapter 4.

Remark 3.5 The numbering of the constraints is according to their occurrence in problem (2.20), e.g. $x_{\text{Se,Ch}} \geq 0$ is constraint number 5.

3.3.1 Interpretation of the Results

In Table 3.1 and Figure 3.5, it is shown that the solution to problem (3.14) is given by four critical regions. This means that for different demand values, there is a change in which of the constraints in problem (3.14) is active.

The first critical region: For the first critical region, the third and fourth constraints are active. This is equivalent to the mp-LP case, as these are the constraints that define the market demand, given from the deterministic values in Eq. (2.17b). Thus, the

Table 3.1 The solution of problem (3.14)

CR_1	$\begin{cases} CR = \left\{\theta \,\middle\|\, \begin{bmatrix} 1 & 0.85 \\ -1 & 0 \\ 0 & -1 \end{bmatrix} \theta \le \begin{bmatrix} 680.56 \\ 0 \\ 0 \end{bmatrix} \right\} \\ k = \{3, 4\} \\ x(\theta) = \begin{bmatrix} 0.51 & 0 \\ 0 & 0.44 \\ 0.49 & 0 \\ 0 & 0.56 \end{bmatrix} \theta \\ \lambda(\theta) = \begin{bmatrix} 272.58 & 0 \\ 0 & 245.52 \end{bmatrix} \theta + \begin{bmatrix} 43.3 \\ 43.3 \end{bmatrix} \end{cases}$
CR_3	$\begin{cases} CR = \left\{\theta \,\middle\|\, \begin{bmatrix} 1 & 1 \\ 1 & -0.78 \\ -1 & 0 \end{bmatrix} \theta \le \begin{bmatrix} 950 \\ -622.22 \\ 0 \end{bmatrix} \right\} \\ k = \{1, 3, 4, 5\} \\ x(\theta) = \begin{bmatrix} 0 & 0 \\ 0 & 0 \\ 1 & 0 \\ 0 & 1 \end{bmatrix} \theta + \begin{bmatrix} 0 \\ 350 \\ 0 \\ -350 \end{bmatrix} \\ \lambda(\theta) = \begin{bmatrix} 0 & 356.38 \\ 561.18 & 0 \\ 0 & 436.38 \\ -324 & 252 \end{bmatrix} \theta + \begin{bmatrix} -2.85 \times 10^5 \\ 43.3 \\ -1.53 \times 10^5 \\ -2.02 \times 10^5 \end{bmatrix} \end{cases}$

(Continued)

Table 3.1 (continued)

CR_2	$\begin{cases} CR = \left\{ \theta \left\| \begin{bmatrix} 1 & 1 \\ -1 & 0.78 \\ 1 & -0.78 \\ -1 & -0.85 \end{bmatrix} \theta \leq \begin{bmatrix} 950 \\ 622.22 \\ 680.56 \\ -680.56 \end{bmatrix} \right. \right\} \\ k = \{1, 3, 4\} \\ x(\theta) = \begin{bmatrix} 0.27 & -0.21 \\ -0.27 & 0.21 \\ 0.73 & 0.21 \\ 0.27 & 0.79 \end{bmatrix} \theta + \begin{bmatrix} 167.16 \\ 182.84 \\ -167.16 \\ -182.84 \end{bmatrix} \\ \lambda(\theta) = \begin{bmatrix} 218.84 & 186.17 \\ 410.42 & 117.26 \\ 117.26 & 345.27 \end{bmatrix} \theta + \begin{bmatrix} -1.49 \times 10^5 \\ -9.38 \times 10^4 \\ -7.98 \times 10^4 \end{bmatrix} \end{cases}$
CR_4	$\begin{cases} CR = \left\{ \theta \left\| \begin{bmatrix} 1 & 1 \\ -1 & 0.78 \\ 0 & -1 \end{bmatrix} \theta \leq \begin{bmatrix} 950 \\ -680.56 \\ 0 \end{bmatrix} \right. \right\} \\ k = \{1, 3, 4, 6\} \\ x(\theta) = \begin{bmatrix} 0 & 0 \\ 0 & 0 \\ 1 & 0 \\ 0 & 1 \end{bmatrix} \theta + \begin{bmatrix} 350 \\ 0 \\ -350 \\ 0 \end{bmatrix} \\ \lambda(\theta) = \begin{bmatrix} 458.21 & 0 \\ 561.18 & 0 \\ 0 & 436.48 \\ 324 & -252 \end{bmatrix} \theta + \begin{bmatrix} -3.12 \times 10^5 \\ -1.96 \times 10^5 \\ 43.3 \\ -2.21 \times 10^5 \end{bmatrix} \end{cases}$

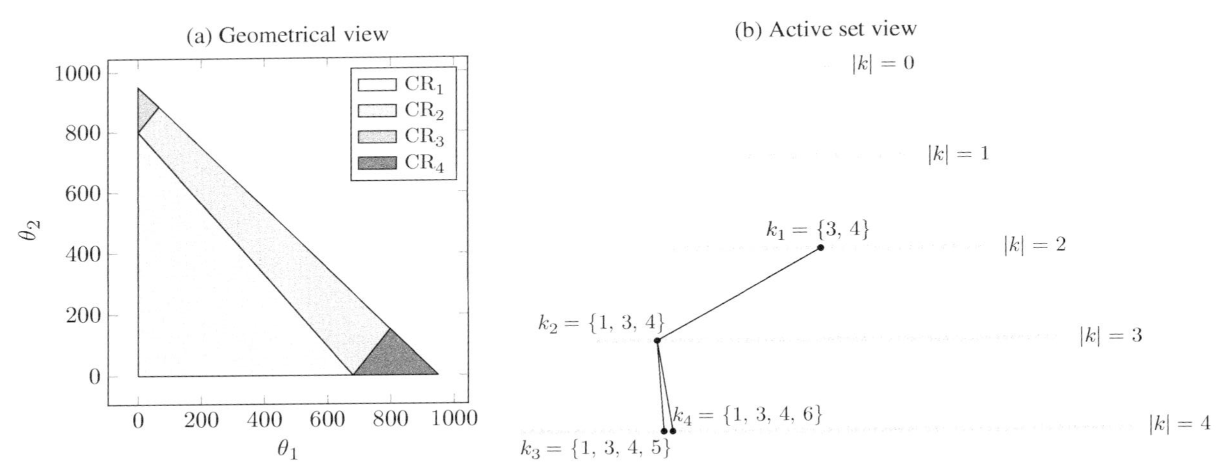

Figure 3.5 The solution to problem (3.14). In (a) the partitioning of the parameter space into four critical regions is shown, each of which is associated with a different set of active constraints k_i. In (b), the gray dots indicate candidate active sets, the number of which depends on the cardinality of the active set, $|k|$. Thus, if $|k| = 1$, then $\binom{8}{1} = 8$ different candidate active sets are possible. Thus, the sum of all points in (b) denotes the maximum number of different active sets for problem (3.14) and consequently the maximum number of critical regions. In addition, the black dots • correspond to the active sets associated with the critical regions shown in (a). In particular, it is evident that the solution is a connected graph and since the conditions of Theorem 5 apply directly, the facet-to-facet property has to hold for the critical regions in (a) as per Lemma 3.1.

solution within the first critical region is only concerned with fulfilling the market demand, as the supply limits are not relevant (see other critical regions). However, on the contrary to the mp-LP case, the optimal solution goes along multiple routes, i.e. the demand for Chicago is satisfied 51% from Seattle and 49% from San Diego ($x_{\text{Se,Ch}} = 0.51\theta_1$ and $x_{\text{SD,Ch}} = 0.49\theta_1$). Conversely, the demand for Topeka is fulfilled 44% from Seattle and 56% from San Diego ($x_{\text{Se,To}} = 0.44\theta_2$ and $x_{\text{SD,To}} = 0.56\theta_2$).

The second critical region: The second critical region is obtained, when the market demand exceeds the available supply of Seattle, which is 350. Concretely, this is given by the first constraint, i.e.

$$0.51\theta_1 + 0.44\theta_2 \leq 350 \quad \Leftrightarrow \quad \theta_1 + 0.85\theta_2 \leq 680.56. \tag{3.15}$$

Note that this is the parametric solution for $x_{\text{Se,Ch}}$ and $x_{\text{Se,To}}$ based on the first critical region. Thus, beyond this constraint, it is necessary to balance the use of the supply from Seattle between Chicago and Topeka. As a result, whenever we satisfy the demand of Chicago from Seattle, we need to reduce the amount shipped to Topeka by the same amount, since we are operating at the maximum capacity, i.e. $x_{\text{Se,Ch}} + x_{\text{Se,To}} = 350$. In numbers, we see that

$$\nabla_\theta \begin{bmatrix} x_{\text{Se,Ch}}(\theta) \\ x_{\text{Se,To}}(\theta) \end{bmatrix} = \begin{bmatrix} 0.27 & -0.21 \\ -0.27 & 0.21 \end{bmatrix}, \tag{3.16}$$

which clearly shows the operation at maximum capacity.

The third critical region: The parametric solution of the second critical region is relevant as long as there is not a single demand that occupies 100% of the supply from one city. However, either with rising demand from Topeka (given by θ_2), discussed in this critical region or from Chicago (given by θ_1), discussed in the next critical region, the supply from Seattle will be allocated entirely to one city. This manifests itself in the fact that the balancing performed for the second critical region would lead to negative values for $x_{\text{Se,Ch}}$, in an effort to counteract the demand from Topeka. In particular, this means that

$$x_{\text{Se,Ch}}(\theta) \leq 0 \quad \Leftrightarrow \quad \theta_1 - 0.78\theta_2 \leq -622.22. \tag{3.17}$$

Once the constraint in Eq. (3.17) becomes active, $x_{\text{Se,Ch}} = 0$ and consequently, as Seattle is still operating at maximum capacity (the first constraint is still active), $x_{\text{Se,To}} = 350$. In addition, the solutions for $x_{\text{SD,Ch}}$ and $x_{\text{SD,To}}$ fulfill the remaining demand from Chicago and Topeka, respectively, i.e. $x_{\text{SD,Ch}} = \theta_1$ and $x_{\text{SD,To}} = \theta_2 - 350$.

The fourth critical region: Similarly to the third critical region, the fourth critical region results when the demand from Chicago causes the supply from Seattle to Topeka to move towards zero. Thus, the non-negativity constraint for $x_{\text{Se,To}}$ becomes active (see the description of the third critical region for the equivalent case for excess demand from Topeka). As a result, $x_{\text{Se,To}} = 0$, $x_{\text{Se,Ch}} = 350$ and the solutions for $x_{\text{SD,Ch}}$ and $x_{\text{SD,To}}$ fulfill the remaining demand from Chicago and Topeka, respectively, i.e. $x_{\text{SD,Ch}} = \theta_1 - 350$ and $x_{\text{SD,To}} = \theta_2$.

Infeasible region: As soon as the sum of the demand, $\theta_1 + \theta_2$, is greater than the available supply, $600 + 350 = 950$, there is no possibility to meet all the demands with the supply given. Thus, there is no feasible solution for $\theta_1 + \theta_2 > 950$.

3.4 Literature Review

It might be assumed that the development of parametric quadratic programming was based on its linear counterpart, or that those developments at least inspired the consideration of such problems. However, on the contrary, parametric quadratic programming in fact pre-dates parametric linear programming, at least on a conceptual level. The first ever description of a parametric quadratic programming problem was given by Harry Markowitz in 1952 in his groundbreaking paper "Portfolio selection," which founded the modern portfolio theory and won him the Nobel Prize in Economics in 1990 [1]. The key idea was to consider the expected return and its variance (which is convex quadratic) of a set of investment possibilities, which naturally is a function of the relative amount invested in each portfolio.[2] In fact, Markowitz

2 Depending on the formulation, it is possible to formulate the problem as a function of the expected return, as done in [2].

states in his paper properties that are still used today such as the continuity of the objective function, which was proven for the general single-parametric case 15 years later by Dantzig *et al.* [3].

Remark 3.6 Note that Markowitz did not develop a general solution procedure in his 1952 paper, but only considered the one- and two-parameter case. In fact, he stated that he "[...] intends to present, in the future, the general, mathematical treatment which removes these limitations," an endeavor he undertook in 1956 [2].

These concepts inspired the classical paper by Wolfe [4] in 1959, where he proposed a method on how to solve (parametric) quadratic programming problems *via* the Simplex algorithm. This opened up the alley these works were the foundation for the different contributions for the single-parameter case [5–10]. In fact, 1976 an entire Ph.D. thesis was dedicated to the study of parametric quadratic programming problems by Jürgen Guddat, an effort that contributed to the classical book by Bank *et al.* on nonlinear parametric optimization in 1983 [11].

However, despite these efforts no publication considering more than one parameter appeared until 2002, where the application of the Basic Sensitivity Theorem founded the body of theory still used today [12, 13] and which enabled the application of multi-parametric programming to model-predictive control. The proposed algorithm was thereby a geometrical algorithm that was based on reversing the constraints of each critical region to ensure the coverage of the parameter space. However, as this leads to artificial cuts, new algorithms were quickly developed, in particular the variable step-size algorithm presented in Chapter 2 [14], as well as geometrical algorithms based on Theorem 3.2 [15–18].

At the same time, several authors considered the idea of exhaustively evaluating all active sets (an approach that was termed "reverse transformation" by Mayne and Raković) [19–23]. However, duc to the combinatorial nature of the problem, the computational complexity was too high to be comparable with the abilities of geometrical algorithms. Only when in 2011 Gupta *et al.* proposed an efficient pruning strategy [24], the computational performance became manageable and gained considerable attention [25, 26].

References

1 Markowitz, H. (1952) Portfolio selection. *The Journal of Finance*, 7 (1), 77–91, doi: 10.1111/j.1540-6261.1952.tb01525.x. URL http://dx.doi.org/10.1111/j.1540-6261.1952.tb01525.x.

2 Markowitz, H. (1956) The optimization of a quadratic function subject to linear constraints. *Naval Research Logistics Quarterly*, 3 (1–2), 111–133, doi: 10.1002/nav.3800030110. URL http://dx.doi.org/10.1002/nav.3800030110.

3 Dantzig, G.B., Folkman, J., and Shapiro, N. (1967) On the continuity of the minimum set of a continuous function. *Journal of Mathematical Analysis and Applications*, 17 (3), 519–548, doi: 10.1016/0022-247X(67)90139-4. URL http://www.sciencedirect.com/science/article/pii/0022247X67901394.

4 Wolfe, P. (1959) The simplex method for quadratic programming. *Econometrica: Journal of the Econometric Society*, 43 (10), 382–398.

5 Ritter, K. (1962) Ein Verfahren zur Losung parameterabhaengiger nichtlinearer Maximumprobleme. *Unternehmensforschung*, 6, 149–196. URL http://download.springer.com/static/pdf/791/art%253A10.1007%252FBF01920852.pdf?auth66=1418818502_e150201592c65d4d944685bb54d0c7a5&ext=.pdf.

6 Boot, J.C.G. (1963) On sensitivity analysis in convex quadratic programming problems. *Operations Research*, 11 (5), 771–786. URL http://www.jstor.org/stable/167911.

7 Kurata, R. (1966) Notes on parametric quadratic programming. *Journal of the Operations Research Society of Japan*, 8 (3), 150–153. URL http://www.orsj.or.jp/~archive/pdf/e_mag/Vol.08_03_150.pdf.

8 Daniel, J.W. (1973) Stability of the solution of definite quadratic programs. *Mathematical Programming*, 5 (1), 41–53, doi: 10.1007/BF01580110. URL http://dx.doi.org/10.1007/BF01580110.

9 Benveniste, R. (1981) One way to solve the parametric quadratic programming problem. *Mathematical Programming*, 21 (1), 224–228, doi: 10.1007/BF01584242. URL http://dx.doi.org/10.1007/BF01584242.

10 Väliaho, H. (1985) A unified approach to one-parametric general quadratic programming. *Mathematical Programming*, 33 (3), 318–338, doi: 10.1007/BF01584380. URL http://dx.doi.org/10.1007/BF01584380.

11 Bank, B., Guddat, J., Klatte, D., Kummer, B., and Tammer, K. (1983) *Non-linear parametric optimization*, Birkhäuser Verlag, Basel and Boston, MA.

12 Bemporad, A., Morari, M., Dua, V., and Pistikopoulos, E.N. (2002) The explicit linear quadratic regulator for constrained systems. *Automatica*, 38 (1), 3–20, doi: 10.1016/S0005-1098(01)00174-1. URL http://www.sciencedirect.com/science/article/pii/S0005109801001741.

13 Dua, V., Bozinis, N.A., and Pistikopoulos, E.N. (2002) A multiparametric programming approach for mixed-integer quadratic engineering problems. *Computers and Chemical Engineering*, 26 (4–5), 715–733, doi: 10.1016/S0098-1354(01)00797-9. URL http://www.sciencedirect.com/science/article/pii/S0098135401007979.

14 Baotic, M. (2002), An efficient algorithm for multi-parametric quadratic programming. Technical Report AUT02-04, Automatic Control Laboratory, ETH Zurich, Switzerland (February 2002). URL https://control.ee.ethz.ch/index.cgi?page=publications;action=details;id=67.

15 Tøndel, P., Johansen, T.A., and Bemporad, A. (2003) An algorithm for multi-parametric quadratic programming and explicit MPC solutions. *Automatica*, 39 (3), 489–497, doi: 10.1016/S0005-1098(02)00250-9. URL http://www.sciencedirect.com/science/article/pii/S0005109802002509.

16 Tøndel, P., Johansen, T., and Bemporad, A. (2003) Further results on multiparametric quadratic programming, in *Proceedings of the 42nd IEEE Conference on Decision and Control, 2003*, vol. 3, pp. 3173–3178, doi: 10.1109/CDC.2003.1273111.

17 Spjøtvold, J., Kerrigan, E.C., Jones, C.N., Tøndel, P., and Johansen, T.A. (2006) On the facet-to-facet property of solutions to convex parametric quadratic programs. *Automatica*, 42 (12), 2209–2214, doi: 10.1016/j.automatica.2006.06.026. URL http://www.sciencedirect.com/science/article/pii/S0005109806002822.

18 Bemporad, A. (2015) A multiparametric quadratic programming algorithm with polyhedral computations based on nonnegative least squares. *IEEE Transactions on Automatic Control*, 60 (11), 2892–2903, doi: 10.1109/TAC.2015.2417851.

19 Seron, M.M., De Dona, J.A., and Goodwin, G.C. (2000) Global analytical model predictive control with input constraints, in *Proceedings of the 39th IEEE Conference on Decision and Control, 2000*, vol. 1, pp. 154–159, doi: 10.1109/CDC.2000.912749.

20 Mayne, D.Q. and Rakovic, S. (2002) Optimal control of constrained piecewise affine discrete time systems using reverse transformation, in *Proceedings of the 41st IEEE Conference on Decision and Control, 2002*, vol. 2, pp. 1546–1551, doi: 10.1109/CDC.2002.1184739.

21 Mayne, D.Q. and Raković, S. (2003) Optimal control of constrained piecewise affine discrete-time systems. *Computational Optimization and Applications*, 25 (1–3), 167–191, doi: 10.1023/A:1022905121198. URL http://dx.doi.org/10.1023/A%3A1022905121198.

22 Munoz de la Pena, D., Alamo, T., Bemporad, A., and Camacho, E.F. (2004) A dynamic programming approach for determining the explicit solution of linear MPC controllers, in *CDC. 43rd IEEE Conference on Decision and Control, 2004*, vol. 3, pp. 2479–2484, doi: 10.1109/CDC.2004.1428785.

23 Seron, M.M., Goodwin, G.C., and De Dona, J.A. (2002) Finitely parameterised implementation of receding horizon control for constrained linear systems, in Proceedings of the 2002 American Control Conference, 2002, vol. 6, pp. 4481–4486, doi: 10.1109/ACC.2002.1025356.

24 Gupta, A., Bhartiya, S., and Nataraj, P. (2011) A novel approach to multiparametric quadratic programming. *Automatica*, 47 (9), 2112–2117, doi: 10.1016/j.automatica.2011.06.019. URL http://www.sciencedirect.com/science/article/pii/S0005109811003190.

25 Feller, C., Johansen, T.A., and Olaru, S. (2013) An improved algorithm for combinatorial multi-parametric quadratic programming. *Automatica*, 49 (5), 1370–1376, doi:

10.1016/j.automatica.2013.02.022. URL http://www.sciencedirect.com/science/article/pii/S0005109813001118.

26 Herceg, M., Jones, C.N., Kvasnica, M., and Morari, M. (2015) Enumeration-based approach to solving parametric linear complementarity problems. *Automatica*, 62, 243–248, doi: 10.1016/j.automatica.2015.09.019. URL http://www.sciencedirect.com/science/article/pii/S0005109815003829.

4

Solution Strategies for mp-LP and mp-QP Problems

In Chapters 2 and 3, mp-LP and mp-QP problems were discussed in detail. In this chapter, the most common algorithms used for the solution of such problems are described. In particular, consider the following mp-QP problem:

$$\begin{aligned} z(\theta) = \quad & \underset{x}{\text{minimize}} && \left(\tfrac{1}{2}Qx + H\theta + c\right)^T x \\ & \text{subject to} && Ax \leq b + F\theta \\ &&& A_{\text{eq}}x = b_{\text{eq}} + F_{\text{eq}}\theta \\ &&& \theta \in \Theta := \{\theta \in \mathbb{R}^q | \text{CR}_A\theta \leq \text{CR}_b\} \\ &&& x \in \mathbb{R}^n, \end{aligned} \tag{4.1}$$

where $Q \in \mathbb{R}^{n\times n}$ is symmetric positive definite, $H \in \mathbb{R}^{n\times q}$, $c \in \mathbb{R}^n$, $A \in \mathbb{R}^{m\times n}$, $b \in \mathbb{R}^m$, $F \in \mathbb{R}^{m\times q}$, $A_{\text{eq}} \in \mathbb{R}^{s\times n}$, $b_{\text{eq}} \in \mathbb{R}^s$, $F_{\text{eq}} \in \mathbb{R}^{s\times q}$, $\text{CR}_A \in \mathbb{R}^{r\times q}$, $\text{CR}_b \in \mathbb{R}^r$, $Ax \leq b$ and θ are compact polytopes, and A_{eq} is assumed to have full rank.[1] Note that problem (4.1) will be used for the general description of the algorithms for mp-LP and mp-QP problems, and the mp-LP case is only discussed separately if appropriate.

In addition, given an active set k, let the active set matrices be defined as

$$A_{\text{AS}} = \left[A_k^T \; A_{\text{eq}}^T\right]^T \tag{4.2a}$$

$$b_{\text{AS}} = \left[b_k^T \; b_{\text{eq}}^T\right]^T \tag{4.2b}$$

1 If A_{eq} does not have full rank, it is always possible to find an equivalent matrix A'_{eq} with a reduced number of rows which has full rank.

Multi-parametric Optimization and Control, First Edition.
Efstratios N. Pistikopoulos, Nikolaos A. Diangelakis, and Richard Oberdieck.

$$F_{\text{AS}} = \left[F_k^T \; F_{\text{eq}}^T\right]^T. \tag{4.2c}$$

Then, the parametric solution for an mp-LP problem is given as

$$\text{CR} = \left\{ \theta \in \mathbb{R}^q \,\middle|\, \begin{bmatrix} A(A_{\text{AS}}^{-1}F_{\text{AS}}) - F \\ \text{CR}_A \end{bmatrix} \theta \leq \begin{bmatrix} b - A(A_{\text{AS}}^{-1}b_{\text{AS}}) \\ \text{CR}_b \end{bmatrix} \right\} \tag{4.3a}$$

$$x(\theta) = A_{\text{AS}}^{-1}(F_{\text{AS}}\theta + b_{\text{AS}}) \tag{4.3b}$$

$$\lambda(\theta) = \lambda_0 \tag{4.3c}$$

$$\mu(\theta) = \mu_0. \tag{4.3d}$$

Equivalently for an mp-QP problem, the parametric solution is given as

$$\text{CR} = \left\{ \theta \in \mathbb{R}^q \,\middle|\, \begin{array}{l} Ax(\theta) \leq b + F\theta \\ \lambda_k(\theta) \geq 0 \\ \text{CR}_A\theta \leq \text{CR}_b \end{array} \right\} \tag{4.4a}$$

$$x(\theta) = -Q^{-1}\left(H\theta + c + \left[A_k^T A_{\text{eq}}^T\right] \begin{bmatrix} \lambda_k \\ \mu \end{bmatrix}\right) \tag{4.4b}$$

$$\begin{bmatrix} \lambda_k(\theta) \\ \mu(\theta) \end{bmatrix} = -(A_{\text{AS}}Q^{-1}A_{\text{AS}}^T)^{-1}(b_{\text{AS}} + F_{\text{AS}}\theta + A_{\text{AS}}Q^{-1}(H\theta + c)). \tag{4.4c}$$

Note that redundant constraints may be present in CR, which can be removed based on the principles discussed in Chapter 1.3.

The rest of the chapter discusses the different algorithms for the full exploration of the parametric space.

4.1 General Overview

Any solution strategy for mp-LP and mp-QP problems needs to consider the following aspects:

- Calculation of the parametric solution and critical region description
- Systematic strategy to search for solutions that have not yet been found
- Convergence in finite time

- Identification of all parametric solutions optimal over the feasible parameter space

Due to the general nature of these requirements, numerous procedures have been presented (see the Literature Review in Chapter 4.6 for a more detailed overview). However, from these procedures three solution approaches have received significant attention as they represent some of the most efficient and general strategies for the solution of mp-LP and mp-QP problems:

Geometrical approach: This approach considers a geometrical interpretation of the problem. In particular, the critical regions are considered as polytopes, and the feasible parameter space is explored by moving in a geometrical fashion from one critical region to another. The different procedures that follow the geometrical approach thereby differ in the way the parameter space is explored and are discussed in Section 4.2.

Combinatorial approach: As shown in Lemma 2.1, every critical region is uniquely defined by its active set. Thus, instead of considering critical regions as polytopes, the combinatorial approach considers all possible combinations of active sets, which are finite, and enumerates each candidate combination. The different algorithms differ in the ability to avoid consideration of all possible combinations and are discussed in Chapter 4.3.

Connected-graph approach: Although the geometrical and combinatorial algorithms solve the same problem, they explore unrelated aspects of the solution properties to obtain the solution. This gap is closed by the connected-graph approach, which utilizes the connected-graph theorem (Theorems 3 and 6, respectively) to integrate geometrical considerations into the combinatorial solution approach. In essence, this means that the number of possible combinations in the combinatorial approach can be drastically reduced due to the consideration of adjacency in a geometrical sense. This approach is discussed in Chapter 4.4.

Remark 4.1 Another avenue of solving mp-LP and mp-QP problems is the reformulation of the original problem into a multi-parametric linear complementarity problem (mp-LCP) through the KKT conditions. For consistency, the discussion of mp-LCP problems is omitted here and can be found in Chapter 9.3.

4.2 The Geometrical Approach

From Eqs. (2.7) and (3.12), it follows that the critical regions resulting from mp-LP and mp-QP problems are polytopes. In particular, each critical region denotes the set of parameters for which the corresponding parametric solution is optimal. Thus, the underlying idea of the geometrical approach is to cover the feasible parameter space with disjoint critical regions, the union of which is convex. Consequently, the following idea was put forth:

(0) Define a starting point θ_0.
(1) Fix θ_0 in problem (4.1), and solve the resulting QP/LP.
(2) Identify the active set for the solution of the QP/LP problem.
(3) Obtain the parametric solution and critical region of the mp-LP or mp-QP problem based on the active set *via* Eqs. (4.3) and (4.4), respectively.
(4) Move outside the found critical region and explore the parameter space. Find a new θ_0, and return to Step 1.

In the following text, each one of these points is discussed in detail.

4.2.1 Define A Starting Point θ_0

In principle, any strategy that enables the automated calculation of a feasible starting point is suitable for the algorithm design. However, in order to ensure the feasibility of problem (4.1), the following statement is required to hold:

$$\mathcal{F} = \left\{ (x, \theta) \;\middle|\; \begin{array}{l} Ax \leq b + F\theta \\ A_{\text{eq}}x = b_{\text{eq}} + F_{\text{eq}}\theta \\ \text{CR}_A\theta \leq \text{CR}_b \end{array} \right\} \neq \emptyset, \tag{4.5}$$

i.e. there exists a set of points (x, θ) such that problem (4.1) is feasible. Consequently, any θ, for which there exists a x such that problem (4.1) is feasible is a valid choice for θ_0. In order to identify such a θ, the solution of the Chebyshev ball is considered (see Chapter 1.3), i.e.

$$\begin{aligned} R = \underset{x,\theta,r}{\text{minimize}} \quad & -r \\ \text{subject to} \quad & K_i \begin{bmatrix} x \\ \theta \end{bmatrix} + r||K_i||_2 \leq b_i, \quad \forall i = 1, \ldots, m, \\ & A_{\text{eq}} = b_{\text{eq}} + F_{\text{eq}}\theta \\ & \text{CR}_{A,i}\theta + r||\text{CR}_{A,i}||_2 \leq \text{CR}_{b,i}, \quad \forall i = 1, \ldots, r, \end{aligned} \tag{4.6}$$

where $K_i = [A_i, -F_i]$. If problem (4.6) is feasible, then θ_0 is set as the corresponding solution for θ.

Remark 4.2 Sometimes, the idea of treating θ as an optimization variable in problem (4.1) is considered as well in the literature. However, as shown in Chapter 3, problem (4.1) may not necessarily be convex in θ. Thus, solver issues could arise and thus problem (4.6) represents a more robust approach.

4.2.2 Fix θ_0 in Problem (4.1), and Solve the Resulting QP

In particular, this transforms problem (4.1) into the following QP:

$$\begin{aligned} z = \underset{x}{\text{Minimize}} \quad & \left(\frac{1}{2}Qx + \tilde{c}\right)^T x \\ \text{Subject to} \quad & Ax \leq \tilde{b} \\ & A_{\text{eq}}x = \tilde{b}_{\text{eq}} \\ & x \in \mathbb{R}^n, \end{aligned} \tag{4.7}$$

where

$$\tilde{c} = c + H\theta_0 \tag{4.8a}$$

$$\tilde{b} = b + F\theta_0 \tag{4.8b}$$

$$\tilde{b}_{\text{eq}} = b_{\text{eq}} + F_{\text{eq}}\theta_0. \tag{4.8c}$$

Since problem (4.7) is a strictly convex QP (or equivalent LP), it can be directly solved with off-the-shelf solvers.

4.2.3 Identify The Active Set for The Solution of The QP Problem

In essence, the active set of the solution of problem (4.7) is given by the following:

$$A_k x_0 = b_k + F_k \theta_0 \tag{4.9a}$$

$$\lambda_{0,k} > 0, \tag{4.9b}$$

where x_0 and λ_0 are the solutions to the QP/LP problem obtained from fixing $\theta = \theta_0$. In the absence of degeneracy, this identification is unique. However, if primal degeneracy is present, then the resulting critical region is lower-dimensional if the active set contains a weakly active constraint (see Chapter 2.2). Thus, in order to identify that situation, the parametric solution for the given active set (using Eq. (4.3) for mp-LP problems and Eq. (4.4) for mp-QP problems is formed and the Chebyshev ball of the resulting critical region is calculated (see problem (1.26)). If the radius of the Chebyshev ball is zero, then the region is lower-dimensional and the corresponding active set can be discarded.

In the case of dual degeneracy, which only can occur in mp-LP problems or mp-QP problems with positive semi-definite Q, the approaches discussed in Chapter 2.2 should be used and to obtain a unique and continuous solution without overlapping critical regions.

4.2.4 Move Outside the Found Critical Region and Explore the Parameter Space

All steps up to this point are common between the different solution procedures, which are based on a geometrical interpretation of the problem. They do differ however in the way the parameter space is explored. The trade-off is thereby between the introduction of artificial cuts and the complete exploration of the parameter space. These different exploration strategies are discussed in the following text, and a corresponding graphical representation is given in Figure 4.1.

Constraint reversal: Given a critical region $CR_0 = \{\theta | G\theta \leq g\}$, the key idea of the constraint reversal strategy is to calculate

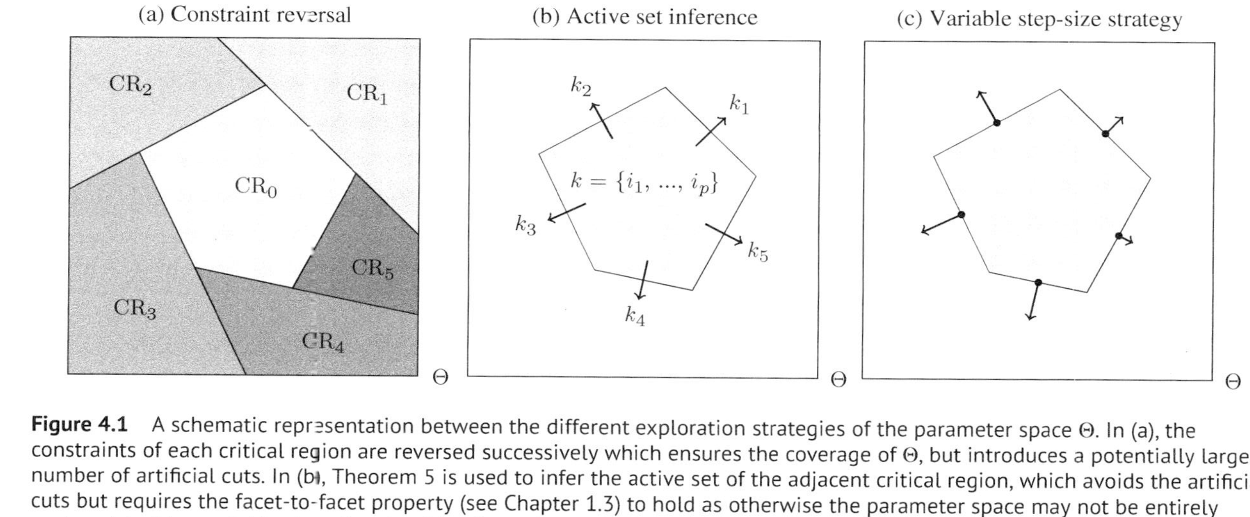

Figure 4.1 A schematic representation between the different exploration strategies of the parameter space Θ. In (a), the constraints of each critical region are reversed successively which ensures the coverage of Θ, but introduces a potentially large number of artificial cuts. In (b), Theorem 5 is used to infer the active set of the adjacent critical region, which avoids the artificial cuts but requires the facet-to-facet property (see Chapter 1.3) to hold as otherwise the parameter space may not be entirely explored. In (c), a variable step-size algorithm is used to move from one critical region to another. While it is a highly efficient algorithm, it is also only guaranteed to find all critical regions if the facet-to-facet property holds for all critical regions.

$\overline{\text{CR}} = \Theta \backslash \text{CR}_0$, i.e. to fully explore the remaining parameter space Θ. This is achieved by successively reversing all the constraints of CR_0 and generating new polytopes CR_i using

$$\text{CR}_i = \left\{ \theta \in \Theta \,\middle|\, \begin{array}{l} G_i\theta > g_i \\ G_j\theta \leq g_j, \forall j < i \end{array} \right\}, \quad i = 1, \ldots, m, \tag{4.10}$$

where m is the total number of constraints in CR_0. For each polytope CR_i, a new value for θ_0 is obtained, e.g. *via* the solution of problem (4.6) for $\theta \in \text{CR}_i$. Note that for any newly obtained solution and corresponding critical region $\text{CR}_{i,0}$, it is necessary to consider $\overline{\text{CR}} = \text{CR}_i \backslash \text{CR}_{i,0}$. Inevitably, this leads to artificial cuts in the parameter space, since two critical regions may feature the same active set but be separated by the generation of CR_i. Thus, from a computational standpoint, it has been shown that this approach is computationally intractable for all but very small example problems. However, it is worth noting that this is the only strategy based on a geometrical approach, which guarantees the exploration of the entire parameter space, regardless of whether the facet-to-facet property holds (see Chapter 1.3), as shown in the following.

The active set inference: Shortly after the description of the constraint reversal approach was presented, Theorem 5 was developed, which paved the way for the design of a strategy based on the inference of the active set of adjacent critical regions. In principle, this approach is very elegant, as it avoids the artificial cuts from the constraint reversal approach. However, as evident from the conditions in Theorem 5, the approach does not enable the unconditional identification of the active set, as it requires the facet-to-facet property to hold. In particular for mp-LP problem, where this property is guaranteed not to hold, this may lead to parts of the parameter space Θ that remain unexplored. In order to tackle this issue, later approaches have combined the idea of the constraint reversal and Theorem 5: if the conditions are proven to hold, then Theorem 5 is utilized, otherwise the constraint reversal is performed. While this does guarantee the exploration of the entire parameter space, it again introduces

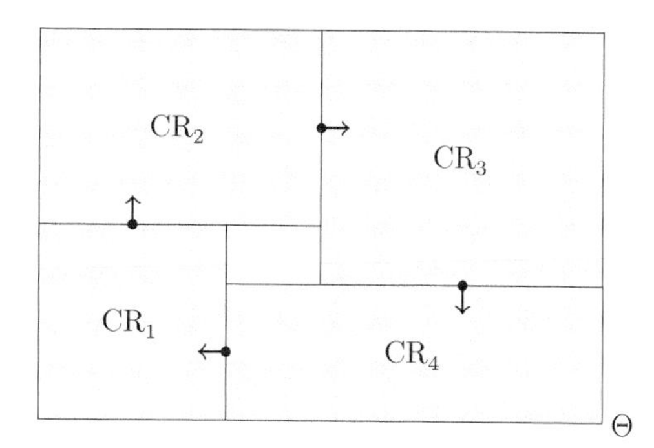

Figure 4.2 A schematic representation of the failure of the variable step-size algorithm to explore the entire parameter space given that the facet-to-facet property does not hold.

artificial cuts. Note that this approach avoids the definition of θ_0 altogether, as it directly jumps to the active set identification step.

The variable step-size approach: The key idea of this approach is based on the idea of exploring each facet of a given critical region CR_0 by finding the "center" of the facet and moving gradually with an increasing step-size orthogonally to the facet beyond a given numerical tolerance. The resulting point θ is used as a new starting point θ_0 for the next iteration. In practice, this idea has been very efficient and many solvers to date still feature an implementation of this solution strategy. However, also this approach is only guaranteed to explore the entire parameter space if the facet-to-facet property holds for all facets, as shown in Figure 4.2.

4.3 The Combinatorial Approach

As stated in Lemma 2.1, every critical region is uniquely defined by its active set. As the number of constraints is finite, it is possible to exhaustively enumerate all possible combinations of active set. This is the key concept behind the combinatorial approach and has been displayed in Figure 3.5b for the example problem of Chapter 3.

Remark 4.3 Some researchers have referred to the combinatorial algorithm also as "reverse transformation" approach, as the problem itself can be transformed into an active set enumeration procedure, the result of which can be transformed in return to yield the solution of the multi-parametric programming problem.

4.3.1 Pruning Criterion

Similarly to the introduction of the artificial cuts in the case of the geometrical algorithm, the direct check of all possible combinations of active sets results in a prohibitively large computational burden for all but small example problems. Thus, the use of the active sets in order to solve multi-parametric programming problems would only be computationally tractable if it was possible to lower the number of candidate active sets considered. Thus, the following pruning criterion was developed:

Lemma 4.1 *Let k_1 and k_2 be two candidate active sets for the solution of problem (4.1), and let $k_1 \subset k_2$. Then, if problem (4.1) is infeasible for k_1, then it will also be infeasible for k_2.*

Proof: Consider the active set k_1 as

$$A_{k_1} x = b_{k_1} + F_{k_1} \theta, \tag{4.11}$$

which restricts the feasible space of problem (3.2) due to the enforced equality. If problem (3.2) subjected to this restriction is infeasible, then an even more restricted problem resulting from $k_2 \supset k_1$ cannot be feasible. □

Consequently, based on Lemma 4.1, it is possible to formulate the following branch-and-bound approach based on an increasing complexity of the active set where the branches associated with an infeasible active set can be discarded (see Figure 4.3):

(0) Set $\mathcal{N} \leftarrow \{\emptyset\}$, $\mathcal{I} \leftarrow \emptyset$, and $\mathcal{S} \leftarrow \emptyset$, where $\mathcal{N}$ denotes the list of active set combinations to be explored, $\mathcal{I}$ contains the active set combinations that have been found to be infeasible, and $\mathcal{S}$ denotes the set of active sets which have been found to result in full-dimensional critical regions.

(1) Get the candidate active set k from $\mathcal{N}$ with the lowest cardinality, i.e. with the lowest number of active constraints. If there are

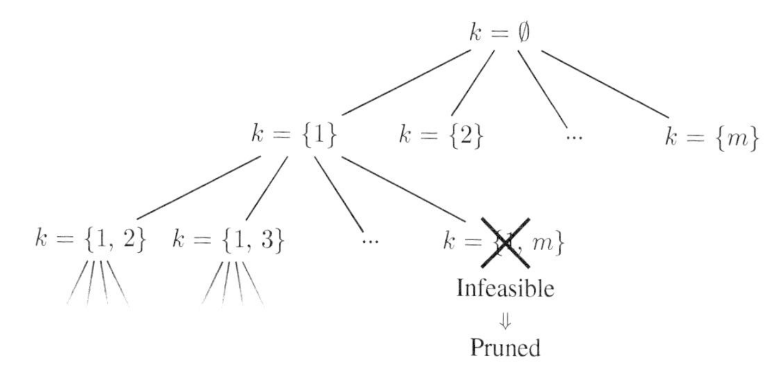

Figure 4.3 A schematic representation of the branch-and-bound algorithm resulting from Lemma 4.1. By successively increasing the complexity of the active set k and discarding the branches associated with infeasible active sets, the solution approach becomes computationally tractable.

no more candidate active sets to explore, i.e. $\mathcal{N} = \emptyset$, the algorithm terminates.

(2) Before verifying the feasibility of the active set k, the following tests are performed:
- Is $k \notin \mathcal{S}$?
- Does $\nexists \tilde{k} \in \mathcal{I}$ such that $\tilde{k} \subset k$?
- Does A_k have full rank?

If one of these conditions is not fulfilled, then k is discarded and the algorithm returns to Step 1.

(3) Let j be the indices of the inactive constraints, i.e. $j = \{1, \ldots, m\} \backslash k$. Then, check whether the considered active set is feasible by verifying the following conditions:

$$\mathcal{F} = \left\{ (x, \theta) \left| \begin{array}{l} A_j x < b_j + F_j \theta \\ A_k x = b_k + F_k \theta \\ A_{\text{eq}} x = b_{\text{eq}} + F_{\text{eq}} \theta \\ \text{CR}_A \theta \leq \text{CR}_b \end{array} \right. \right\} \neq \emptyset. \tag{4.12}$$

If the condition in Eq. (4.12) holds, i.e. if $\mathcal{F} \neq \emptyset$, then there exists a pair (x, θ), which is feasible and the algorithm proceeds to the next step. However, if $\mathcal{F} = \emptyset$, then $\mathcal{I} \leftarrow k$ and the algorithm returns to Step 1.

(4) Calculate the parametric solution as per Eqs. (4.3) and (4.4), respectively.

(5) Check whether the resulting critical region is full-dimensional *via* the solution of the Chebyshev ball. If the critical region is full-dimensional, set $S \leftarrow k$.

(6) If the cardinality of k is less than the size of the optimization variable,[2] i.e. $|k| < n$, then a series of new candidate active sets is generated according to the following:

$$k_i = k \cup \{j_i\}, \quad \forall i = 1, \dots, |j|, \tag{4.13}$$

where $j = \{1, \dots, m\} \backslash k$ are the indices of the inactive constraints. At this point, $\mathcal{N} \leftarrow k_i, \forall i$ and the algorithm returns to Step 1.

Remark 4.4 The benefit of Lemma 4.1 is based on the ability to transform the exhaustive enumeration into a branch-and-bound approach. However, in the case of mp-LP problems, the cardinality of the active set is fixed to the dimensions of the optimization variable, i.e. $|k| = n$. Thus, even if Lemma 4.1 reduces the number of candidate active sets, all the solutions will still be found on the last level of the branch and bound tree.

4.4 The Connected-Graph Approach

The main advantage of the geometrical algorithm is the intuitive use of the critical region as well as the utilization of the concept of adjacent critical regions. Conversely, the combinatorial algorithm completely avoids the need for the facet-to-facet property to hold while guaranteeing that all optimal active sets are found and the entire parameter space is explored. Thus, although both algorithms solve the same problem, they explore unrelated solution properties.

The main idea behind the connected-graph approach is to introduce the concept of adjacent critical regions into the combinatorial setting. In particular, it reduces the number of candidate active sets to be considered in the combinatorial algorithm[3] by

2 This is equivalent to stating that the active constraints do not form a vertex.
3 Thus, the connected-graph approach could be viewed as the introduction of an additional pruning criterion.

considering which candidate active sets could be adjacent to an active set, which results in a full-dimensional critical region.

The algorithm is thereby subdivided into the following steps:

(0) In order to initialize the algorithm, a first active set k_0 that yields a full-dimensional critical region needs to be found. This may be achieved in two ways:
   - Utilize the combinatorial algorithm, until the first active set has been found.
   - Perform the first iteration of the geometrical algorithm.

   After k_0 has been identified, the sets $\mathcal{N} \leftarrow k_0$, $\mathcal{S}$, and $\mathcal{I}$ are defined analogously to the combinatorial algorithm.[4]

(1) If $\mathcal{N} = \emptyset$, then the algorithm terminates. Otherwise, a candidate active set k with the lowest cardinality is chosen from $\mathcal{N}$, and, analogously to the combinatorial algorithm, the following tests are performed:
   - Is $k \notin \mathcal{S}$?
   - Does $\nexists \tilde{k} \in \mathcal{I}$ such that $\tilde{k} \subset k$?
   - Does A_k have full rank?

   If one of these conditions is not fulfilled, then k is discarded and a new candidate active set k is selected.

(2) Let j be the indices of the inactive constraints, i.e. $j = \{1, \ldots, m\} \backslash k$. Then, check whether the considered active set is feasible by verifying the following conditions:

$$\mathcal{F} = \left\{ (x, \theta) \left| \begin{array}{l} A_j x \leq b_j + F_j \theta \\ A_k x = b_k + F_k \theta \\ A_{\mathrm{eq}} x = b_{\mathrm{eq}} + F_{\mathrm{eq}} \theta \\ \mathrm{CR}_A \theta \leq \mathrm{CR}_b \end{array} \right. \right\} \neq \emptyset. \tag{4.14}$$

   If the condition in Eq. (4.12) holds, i.e. if $\mathcal{F} \neq \emptyset$, then there exists a pair (x, θ), which is feasible and the algorithm proceeds to the next step. However, if $\mathcal{F} = \emptyset$, then $\mathcal{I} \leftarrow k$ and the algorithm returns to Step 1.

(3) Calculate the parametric solution as per Eqs. (4.3) and (4.4), respectively. In addition, the minimal representation of

---

4 Note that the set k_0 is not directly added to $\mathcal{S}$ in order to facilitate the description of the algorithm.

the critical region is obtained by removing all redundant constraints according to the principles in Chapter 1.3.

(4) As a result of the removal of redundant constraints, the following conclusions for the resulting critical region CR can be drawn:

CR $= \emptyset$: Based on the feasibility check in Step 2, CR $= \emptyset$ indicates that the parametric solution is not optimal, as the only constraint that is added beyond the constraints in Eq. (4.14) are $\lambda(\theta) \geq 0$, which ensure the optimality of the solution. Thus, no critical region is formed and the algorithm returns to Step 1.

CR *is lower-dimensional*: This situation occurs in the case of primal degeneracy (see Chapter 2.2). Thus, one or more of the elements of the active set k need to be removed in order to obtain a full-dimensional critical region. Thus, the following candidates are generated:

$$\mathcal{L} = \binom{k}{|k|-1}, \tag{4.15}$$

and are added to $\mathcal{N}$.

CR *is full-dimensional*: If the critical region is full-dimensional, each facet of CR_k can be classified into Type I or II from Theorem 5, or as the borders of Θ. In particular, this means that the constraint, which forms the facet of CR is classified as follows:

$$Ax(\theta) \leq b + F\theta \qquad \text{(Type I)} \tag{4.16a}$$

$$\lambda_k(\theta) \geq 0 \qquad \text{(Type II)} \tag{4.16b}$$

$$\mathrm{CR}_A\theta \leq \mathrm{CR}_b \qquad \text{(Border of } \Theta\text{)} \tag{4.16c}$$

If the facet is a border of Θ, there cannot be any adjacent region. If the facet is of Type II, then it is clear that LICQ will hold, since LICQ holds for k as full rank was established in Step 1. Thus, the assumptions from Theorem 5 are directly fulfilled,[5] the facet-to-facet property holds and the active set of the adjacent critical region is added to the set of candidate active sets $\mathcal{N}$. Conversely, if the facet is of Type I, then let k_+ denote the

5 Note that Type II constraints are absent from critical regions resulting in mp-LP problems, as the Lagrange multipliers are constant, i.e. $\lambda(\theta) = \lambda_0$.

active set obtained from Theorem 5. If A_{k_+} has full rank, then Theorem 5 applies and the active set of the adjacent critical region is added to the set of candidate active sets $\mathcal{N}$. However, if full rank is not established, then similarly to the case of a lower-dimensional CR full rank can only be established by removing one of the elements of k_+, i.e. to generate the following candidates:

$$\mathcal{L} = \binom{k_+}{|k|}, \tag{4.17}$$

and to add $\mathcal{L}$ to $\mathcal{N}$. Note that this corresponds to one step of the dual simplex algorithm.

Remark 4.5 In the case where the critical region CR features two or more identical constraints, i.e. $\exists i, j$ such that $A_i\theta - b_i = A_j\theta - b_j$ for all θ, the indices of all identical constraints corresponding to facets of the critical region are considered.

At this point, the algorithm returns to Step 1.

Remark 4.6 In the case of dual degeneracy, disconnected graphs may result. Unfortunately, the principles described in Chapter 2.2 have only partially been applied to the connected graph algorithm, as Gal and Nedoma discussed a strategy for avoiding overlapping critical regions resulting from dual degeneracy [1].

4.5 Discussion

None of the algorithms presented in this chapter have been proven to improve on the worst-case complexity of solving problem (4.1). As a result, any comparison between them highly depends on the specific problem under consideration. However, several heuristics have been developed, which mirror the strengths and weaknesses of the different algorithms.

Remark 4.7 Note however that these are merely indicative and it is still impossible to predict prior to solving the problem which approach – geometrical, combinatorial, or connected-graph – will be the most efficient technique.

The geometrical approach:

- In the case of well-behaved mp-QP problems, it is highly efficient.
- For pathological mp-QP problems and in general for mp-LP problems, the risk of incomplete exploration of the parameter space is significant and thus the use of a geometrical approach should be avoided.
- The algorithm tends to scale well if the number of optimization variables increases.
- For problems with large numbers of constraints, the algorithm tends to perform poorly due to the requirement of removing redundant constraints at each step.

The combinatorial approach:

- The combinatorial algorithm is ill-suited for mp-LP problems as it requires the exhaustive enumeration of all options.
- For problems with a large number of constraints but few optimization variables, the combinatorial algorithm has been proven to be effective.
- If the problem contains symmetry elements, then this can be utilized to increase the pruning efficiency as thus the overall efficiency of the algorithm.

The connected-graph approach:

- As the connected graph approach is guaranteed to explore the entire parameter space, it is well suited for the solution of mp-LP problems as it does not necessarily require the exhaustive enumeration of all combinations.
- The connected-graph approach is ill-suited for dual degenerate mp-LP problems, as the presence of disconnected graphs may result in incomplete parameter space exploration.
- In the case of well-behaved mp-QP problems, the connected-graph approach has also been shown to be highly efficient since, if Theorem 5 is applicable, the adjacent active set can be identified unambiguously, thus dramatically reducing the number of candidate active sets to consider.

- Similarly to the geometrical approach, the connected-graph approach relies on the removal of redundant constraints. Thus, problems featuring large number of constraints tend to be ill-suited for the connected-graph approach.

4.6 Literature Review

As may be expected, many researchers who have worked on multi-parametric programming have done so by proposing a novel solution technique, or by applying an existing solution technique to new problems. Thus, the number of algorithms that have been presented in the literature to solve mp-LP and mp-QP problems is quite significant. In particular, the material presented in this chapter only represents those algorithms that have become very popular due to their applicability, ease of implementation and favorable performance.

As mentioned in Chapters 2 and 3, up until the 1990s, it was quite rare to consider parametric programming problems featuring multiple parameters. While Dinkelbach did propose a solution method for parametric programming problems featuring two parameters based on a geometrical approach [2], the most meaningful algorithm, a connected-graph approach, was presented in 1972 by Gal and Nedoma [1]. The historical development during this period has been reported on extensively in [3].

Remark 4.8 It is very important to note that the algorithm by Gal and Nedoma only represents one of the many algorithms presented between the 1950s and the 1990s. In particular, the group out of Berlin, Germany, around Prof. Nožička made tremendous contributions to the area of general parametric programming problems, and even considered the case of general multi-parametric non-linear programming. This resulted in the excellent textbook by Bank *et al.* [4], which summarized most of these developments. However, most of the solution approaches presented therein are of limited practical applicability and have not had the same impact as the strategies presented in this chapter.

After the development of the exact solution of mp-QP problems based on the application of the basic sensitivity theorem, the first general geometrical algorithm based on constraint reversal was presented in 2000 [5–7], while in 2002, Baotić described the variable step-size approach in a Technical Report [8], and in 2003 Tøndel *et al.* uncovered Theorem 5 and described a geometrical approach based on the inference of the active set. However, it was not until the work by Spjøtvold *et al.* in 2006 that the impact of the facet-to-facet property became apparent [9, 10], which led to the description the geometrical approach featuring the active set inference and the variable step-size approach.

Conversely, the idea of exhaustively enumerating all active sets was proposed in 2001–2003 by Mayne and Raković [11–13] and termed "reverse transformation." However, due to its computational limitation, it was not until in 2011 Gupta *et al.* proved Lemma 4.1 and paved the way for the description of the combinatorial algorithm shown in this Chapter [14]. Building upon this contribution, Feller and Johansen have shown that the use of symmetry can lead to significant improvements in the computational behavior [15, 16].

After the use of the connected-graph Theorem 3 by Gal and Nedoma [1], no other contribution had further developed this area of research until in 2016 the authors of this book were able to generalize Theorem 3 to the mp-QP case and thus prove Theorem 6 and design the connected-graph algorithm presented in this Chapter [17].

As previously mentioned, several algorithms have been presented for the solution of mp-LP and mp-QP problems, two of which are briefly discussed in the succeeding text. First, in 2010 Patrinos and Sarimveis utilized the concepts of graphical derivatives to obtain the solution of an mp-QP problem [18]. In essence, this approach parametrizes the gradient search and thus enables the exploration of a parameter space. Second, in 2012 Mönnigmann and Jost described in the context of explicit model predictive control a strategy where the vertices of the critical region are used to derive the solution of mp-QP problem *via* interpolation.

References

1 Gal, T. and Nedoma, J. (1972) Multiparametric linear programming. *Management Science*, 18 (7), 406–422, doi: 10.1287/mnsc.18.7.406.

2 Dinkelbach, W. (1969) *Sensitivitätsanalysen und parametrische Programmierung, Ökonometrie und Unternehmensforschung / Econometrics and Operations Research*, vol. 12, Springer-Verlag, Berlin, Heidelberg.

3 Gal, T. (1985) The historical development of parametric programming, in *Parametric optimization and approximation, International series of numerical mathematics / Internationale Schriftenreihe zur Numerischen Mathematik / Série internationale d'Analyse numérique*, vol. 72 (eds B. Brosowski and F. Deutsch), Birkhäuser, Basel, pp. 148–165, doi: 10.1007/978-3-0348-6253-0.10. URL http://dx.doi.org/10.1007/978-3-0348-6253-0. 10.

4 Bank, B., Guddat, J., Klatte, D., Kummer, B., and Tammer, K. (1983) *Non-linear parametric optimization*, Birkhäuser Verlag, Basel and Boston, MA.

5 Bemporad, A., Borrelli, F., and Morari, M. (2000) Piecewise linear optimal controllers for hybrid systems, in *Proceedings of the American Control Conference*, vol. 2, pp. 1190–1194, doi: 10.1109/ACC.2000.876688.

6 Pistikopoulos, E.N., Dua, V., Bozinis, N.A., Bemporad, A., and Morari, M. (2000) On-line optimization via off-line parametric optimization tools. *Computers and Chemical Engineering*, 24 (2–7), 183–188, doi: 10.1016/S0098-1354(00)00510-X. URL http://www.sciencedirect.com/science/article/pii/S009813540000510X.

7 Bemporad, A., Morari, M., Dua, V., and Pistikopoulos, E.N. (2002) The explicit linear quadratic regulator for constrained systems. *Automatica*, 38 (1), 3–20, doi: 10.1016/S0005-1098(01)00174-1. URL http://www.sciencedirect.com/science/article/pii/S0005109801001741.

8 Baotic, M. (2002) *An efficient algorithm for multi-parametric quadratic programming*. Technical Report AUT02-04, Automatic

Control Laboratory, ETH Zurich, Switzerland (February 2002). URL https://control.ee.ethz.ch/index.cgi?page=publications;action=details;id=67.

9 Spjötvold, J., Kerrigan, E.C., Jones, C.N., Töndel, P., and Johansen, T.A. (2006) On the facet-to-facet property of solutions to convex parametric quadratic programs. *Automatica*, 42 (12), 2209–2214, doi: 10.1016/j.automatica.2006.06.026. URL http://www.sciencedirect.com/science/article/pii/S0005109806002822.

10 Spjotvold, J., Kerrigan, E.C., Jones, C.N., Tondel, P., and Johansen, T.A. (2006) On the facet-to-facet property of solutions to convex parametric quadratic programs, in *Mathematical Theory of Networks and Systems*, Kyoto, Japan. URL http://control.ee.ethz.ch/index.cgi?page=publications;action=details;id=2696.

11 Mayne, D.Q. (2001) Control of constrained dynamic systems. *European Journal of Control*, 7 (2–3), 87–99, doi: 10.3166/ejc.7.87-99. URL http://www.sciencedirect.com/science/article/pii/S0947358001711417.

12 Mayne, D.Q. and Rakovic, S. (2002) Optimal control of constrained piecewise affine discrete time systems using reverse transformation, in *Proceedings of the 41st IEEE Conference on Decision and Control, 2002*, vol. 2, pp. 1546–1551, doi: 10.1109/CDC.2002.1184739.

13 Mayne, D.Q. and Raković, S. (2003) Optimal control of constrained piecewise affine discrete-time systems. *Computational Optimization and Applications*, 25 (1–3), 167–191, doi: 10.1023/A:1022905121198. URL http://dx.doi.org/10.1023/A%3A1022905121198.

14 Gupta, A., Bhartiya, S., and Nataraj, P. (2011) A novel approach to multiparametric quadratic programming. *Automatica*, 47 (9), 2112–2117, doi: 10.1016/j.automatica.2011.06.019. URL http://www.sciencedirect.com/science/article/pii/S0005109811003190.

15 Feller, C., Johansen, T.A., and Olaru, S. (2013) An improved algorithm for combinatorial multi-parametric quadratic programming. *Automatica*, 49 (5), 1370–1376, doi: 10.1016/j.automatica.2013.02.022. URL http://www.sciencedirect.com/science/article/pii/S0005109813001118.

16 Feller, C. and Johansen, T.A. (2013) Explicit MPC of higher-order linear processes via combinatorial multi-parametric quadratic programming, in *2013 European Control Conference (ECC)*, pp. 536–541.

17 Oberdieck, R., Diangelakis, N.A., and Pistikopoulos, E.N. (2017) Explicit model predictive control: a connected-graph approach. *Automatica*, 76, 103–112, ISSN 0005-1098, doi: 10.1016/j.automatica.2016.10.005.

18 Patrinos, P. and Sarimveis, H. (2010) A new algorithm for solving convex parametric quadratic programs based on graphical derivatives of solution mappings. *Automatica*, 46 (9), 1405–1418, doi: 10.1016/j.automatica.2010.06.008. URL http://www.sciencedirect.com/science/article/pii/S000510981000258X.

5

Multi-parametric Mixed-integer Linear Programming

Consider the following mixed-integer linear programming (MILP) problem:

$$
\begin{aligned}
z = \quad & \underset{x,y}{\text{Minimize}} && c^T\omega \\
& \text{Subject to} && Ax + Ey \leq b \\
& && A_{\text{eq}}x + E_{\text{eq}}y = b_{\text{eq}} \\
& && x \in \mathbb{R}^n, \ \ y \in \{0,1\}^p, \ \ \omega = [x^T \ y^T]^T,
\end{aligned}
\tag{5.1}
$$

where $c \in \mathbb{R}^{(n+p)}$, $A \in \mathbb{R}^{m\times n}$, $E \in \mathbb{R}^{m\times p}$, $b \in \mathbb{R}^m$, $A_{\text{eq}} \in \mathbb{R}^{s\times n}$, $E_{\text{eq}}\mathbb{R}^{s\times p}$, $b_{\text{eq}} \in \mathbb{R}^s$ and $Ax \leq (b - E\overline{y})$ is a compact polytope for all feasible $\overline{y} \in \{0,1\}^p$. The ability to handle optimization problems featuring binary variables is crucial to any situation where discrete elements are present, such as in design, scheduling, and control of hybrid systems. The key difficulty thereby arises from the disjoint and thus non-convex nature inherent to MILP problems. However, even in such problems values such as prices, demand, and risk might change over time and have great impact on the solution. In order to address this, the uncertainty can be considered explicitly by formulating a multi-parametric mixed-integer linear programming (mp-MILP) problem:

$$
\begin{aligned}
z(\theta) = \quad & \underset{x,y}{\text{Minimize}} && c^T\omega \\
& \text{Subject to} && Ax + Ey \leq b + F\theta \\
& && A_{\text{eq}}x + E_{\text{eq}}y = b_{\text{eq}} + F_{\text{eq}}\theta \\
& && \theta \in \Theta := \{\theta \in \mathbb{R}^q | \text{CR}_A\theta \leq \text{CR}_b\} \\
& && x \in \mathbb{R}^n, \ \ y \in \{0,1\}^p, \ \ \omega = [x^T \ y^T]^T,
\end{aligned}
\tag{5.2}
$$

Multi-parametric Optimization and Control, First Edition.
Efstratios N. Pistikopoulos, Nikolaos A. Diangelakis, and Richard Oberdieck.

where $F \in \mathbb{R}^{m \times q}$, $F_{\text{eq}} \in \mathbb{R}^{s \times q}$, $\text{CR}_A \in \mathbb{R}^{r \times q}$, $\text{CR}_b \in \mathbb{R}^r$ and Θ is a compact polytope. Note that it is assumed that there do not exist any equality constraints featuring only the parameters, as this would render the parameter space lower-dimensional. The difference between problems (5.1) and (5.2) is thereby the presence of the bounded uncertain parameters θ in the objective function and the constraints, which are not considered as optimization variables. As a result, the solution x and the optimal objective function z of problem (5.2) are obtained as a function of θ, i.e. $x(\theta)$ and $z(\theta)$, respectively. Note that it is possible to add a scalar d to the objective function of the MILP problem in (5.1) without influencing the optimal solution. Similarly, it is possible to add an arbitrary scaling function $f(\theta)$ to the mp-MILP in (5.2) problem without influencing the optimal solution.

Remark 5.1 Although the term "mixed-integer" would indicate that general integer variables are considered, i.e. $d = \{n_{d_0}, n_{d_0} + 1, \dots, n_{d_1}\}^p$, in this chapter we only consider binary variables. Note however that it is possible to model any integer variable d as a set of $(d_0 + d_1) + 1$ binary variables, $y_i \in \{0, 1\}$, via

$$\sum_{i=n_{d_0}}^{n_{d_1}} i y_i = d \tag{5.3a}$$

$$\sum_{i=n_{d_0}}^{n_{d_1}} y_i = 1. \tag{5.3b}$$

5.1 Solution Properties

5.1.1 From mp-LP to mp-MILP Problems

Conceptually, the presence of binary variables can be understood as a set of discrete choices, each of which can be selected by fixing the binary variables to the corresponding value, resulting in an mp-LP problem of type (2.2). Thus, problem (5.2) can be regarded as 2^p different mp-LP problems. As a result, it is clear that the solution properties of mp-MILP problems will consist of the solution properties

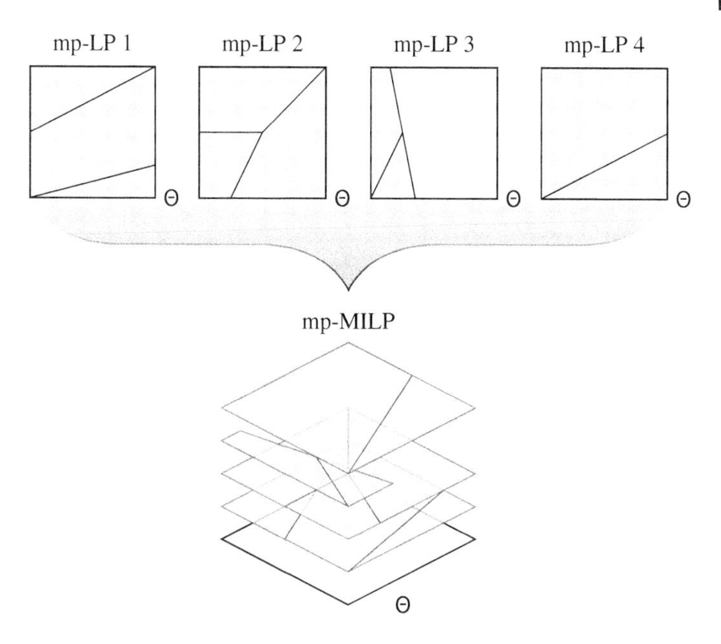

Figure 5.1 The interpretation of mp-MILP problems as a combination of several mp-LP problems.

of mp-LP problems combined with the changes incurred by considering more than one solution in the same parameter space. This concept is schematically depicted in Figure 5.1.

5.1.2 The Properties

This leads to the following theorem:

Theorem 5.1 ***(The solution of mp-MILP problems)*** *Consider the mp-MILP problem (5.2). Then the set of feasible parameters $\Theta_f \subseteq \Theta$ is a possibly non-convex union of polytopes, and the optimizer $x(\theta) : \Theta_f \mapsto \mathbb{R}^n$ and the optimal objective function $z(\theta) : \Theta_f \mapsto \mathbb{R}$ are piecewise affine.*

Proof: The solution of an mp-MILP problem can be considered the union of the solution of several mp-LP problems. Thus, the key thing to prove is that the optimal solution when comparing the solution of several mp-LP problems is a set of polytopes featuring piecewise affine optimizers and solutions. Since each optimal solution corresponds to the solution of an mp-LP problem, it is clear that the optimizer and the optimal objective function are piecewise affine, as this property originates from the solution of the mp-LP problem. Thus, the only thing to show that the transition between the different solutions is affine, i.e.

$$\Delta z(\theta) = z_1(\theta) - z_2(\theta) = 0, \tag{5.4}$$

where $\Delta z(\theta)$ is the transition between the solution of mp-LP problems 1 and 2 with the optimal objective functions $z_1(\theta)$ and $z_2(\theta)$, respectively. Since $z_1(\theta)$ and $z_2(\theta)$ are affine, so is Δz and thus the partitioning of the mp-MILP problem is polytopic. □

5.2 Comparing the Solutions from Different mp-LP Problems

Remark 5.2 Due to the similarities between the solution procedures for mp-MILP and mp-MIQP problems, they are described together in Chapter 7.

From Theorem 5.1 and Figure 5.1, it is clear that the key operation for mp-MILP problems is the comparison of several parametric solutions. The solution of an mp-LP problem of type (2.2) is given by its feasible parameter space Θ_f partitioned into critical regions CR_i, associated solutions $x_i(\theta)$, and optimal objective functions $z_i(\theta)$. In particular, it is clear that the solution that results in the lower objective function value at a given parameter realization θ is considered optimal. An example of a comparison procedure is schematically shown in Figure 5.2.

Thus, in order to compare the solutions from two mp-LP problems,[1] the following operations need to be performed:

---

1 Once the tools to compare the solutions from two mp-LP problems have been developed, it is straightforward to extend them to the case of n mp-LP problems.

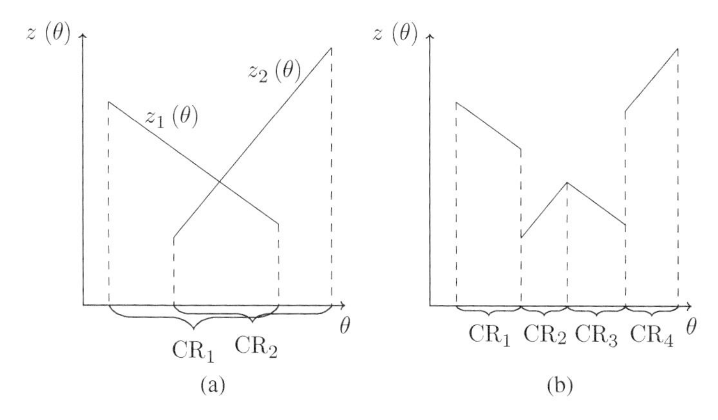

Figure 5.2 The schematic representation of the example of a comparison procedure. In (a), the optimal objective function values of a critical region from two mp-LP problems are shown as a function of the parameter. As they overlap, (b) shows the resulting optimal objective function value resulting from the comparison of the two optimal objective functions for the critical regions, over which they are defined. Note that the choice of which solution is optimal is purely driven by the need to obtain a lower objective function value.

Remark 5.3 In order to distinguish the two solutions, all elements associated with the solution of the second mp-LP problem are denoted with a bar, i.e. $\bar{\cdot}$.

5.2.1 Identification of Overlapping Critical Regions

In order to compare two optimal objective functions, $z_i(\theta)$ and $\bar{z}_j(\theta)$, they need to be valid in the same region of the parameter space. Thus, every combination of CR_i and $\overline{\text{CR}}_j$ is classified into one of the following cases (see Figure 5.3):

- *Case* 1: $\widetilde{\text{CR}}_{ij} = \overline{\text{CR}}_j \cap \text{CR}_i = \overline{\text{CR}}_j$.
- *Case* 2: $\widetilde{\text{CR}}_{ij} = \overline{\text{CR}}_j \cap \text{CR}_i = \text{CR}_i$.
- *Case* 3: $\widetilde{\text{CR}}_{ij} = \overline{\text{CR}}_j \cap \text{CR}_i \neq \emptyset$.
- *Case* 4: $\widetilde{\text{CR}}_{ij} = \overline{\text{CR}}_j \cap \text{CR}_i = \emptyset$.

If $\widetilde{\text{CR}}_{ij} \neq \emptyset$ (i.e. Cases 1–3), then the two solutions need to be compared against each other.

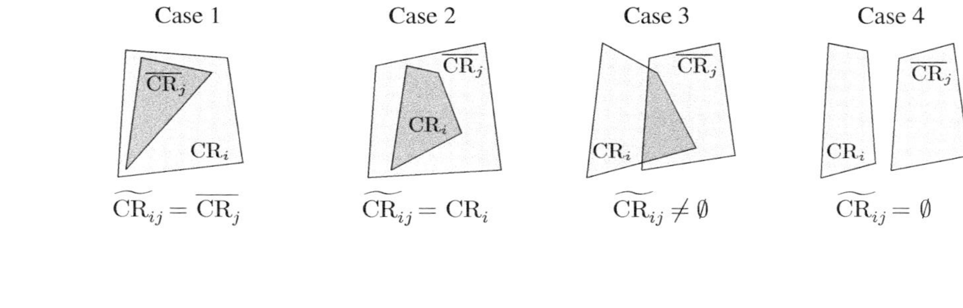

Figure 5.3 The detection of the intersection between two critical regions, $\widetilde{\mathrm{CR}}_{ij} = \overline{\mathrm{CR}}_j \cap \mathrm{CR}_i$.

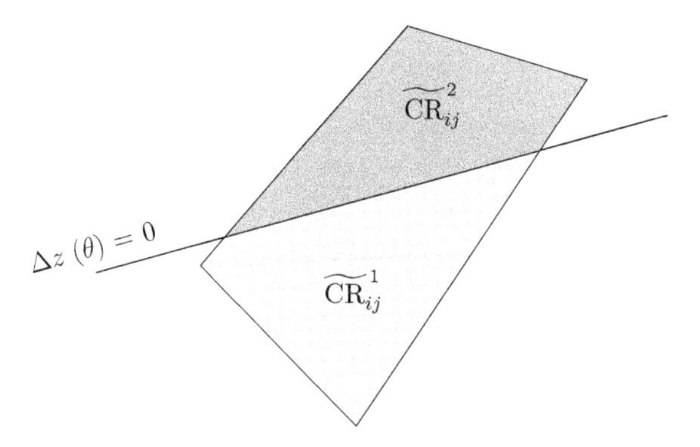

Figure 5.4 The partitioning of $\widetilde{\mathrm{CR}}_{ij}$ into $\widetilde{\mathrm{CR}}_{ij}^1$ and $\widetilde{\mathrm{CR}}_{ij}^2$ based on Eq. (5.7).

5.2.2 Performing the Comparison

Once an overlap $\widetilde{\mathrm{CR}}_{ij}$ has been identified, the comparison between the two solutions can be expressed as

$$z(\theta) = \min\ \{\overline{z}_j(\theta), z_i(\theta)\}, \tag{5.5}$$

which directly leads to the following difference equation:

$$\Delta z(\theta) = \overline{z}_j(\theta) - z_i(\theta) \overset{!}{=} 0. \tag{5.6}$$

As a result, the solution of Eq. (5.5) is a partitioning of $\widetilde{\mathrm{CR}}_{ij}$ according to (see Figure 5.4):

$$\widetilde{\mathrm{CR}}_{ij}^1 = \widetilde{\mathrm{CR}}_{ij} \cap \Delta z(\theta) \leq 0 \qquad \Rightarrow \overline{z}_j(\theta) \text{ optimal} \tag{5.7a}$$

$$\widetilde{\mathrm{CR}}_{ij}^2 = \widetilde{\mathrm{CR}}_{ij} \cap \Delta z(\theta) \geq 0 \qquad \Rightarrow z_i(\theta) \text{ optimal.} \tag{5.7b}$$

Note that since $\overline{z}_j(\theta)$ and $z_i(\theta)$ are affine, $\widetilde{\mathrm{CR}}_{ij}^1$ and $\widetilde{\mathrm{CR}}_{ij}^2$ are polytopes.

5.2.3 Constraint Reversal for Coverage of Parameter Space

As schematically represented in Figure 5.2, the space outside of the overlapping parameter space $\widetilde{\mathrm{CR}}_{ij}$ also needs to be considered. In particular, consider the critical region CR_i: in the Cases 2 and 4 (see Figure 5.3), the entire critical region has been considered. However, the Cases 1 and 3 result in a partial consideration of CR_i, i.e. only $\widetilde{\mathrm{CR}}_{ij} \subset \mathrm{CR}_i$ has been considered. As a result, it is necessary to retain the remaining part of CR_i, i.e. $\mathrm{CR}_i \backslash \widetilde{\mathrm{CR}}_{ij}$. In essence, there are two ways to calculate $\mathrm{CR}_i \backslash \widetilde{\mathrm{CR}}_{ij}$: either via constraint reversal or modelling using integer variables (see Chapter 1.3). In order not to increase the complexity of the underlying problem with the addition of new binary variables, the approach of constraint reversal is used, i.e. let $\mathrm{CR}_i = \{\theta | G\theta \leq g\}$:

$$\mathrm{CR}_i \backslash \widetilde{\mathrm{CR}}_{ij} = \left\{ \theta \in \mathrm{CR}_i \;\middle|\; \begin{array}{l} G_i\theta > g_i \\ G_j\theta \leq g_j, \;\; \forall j < i \end{array} \right\}, \quad i = 1, \dots, m. \tag{5.8}$$

This operation ensures that the entire region of CR_i is taken into consideration.

5.3 Multi-parametric Integer Linear Programming

Consider problem (5.2) with $n = 0$, i.e.:

$$\begin{aligned} z(\theta) = \quad & \underset{y}{\text{Minimize}} \quad c^T y \\ & \text{Subject to} \quad Ey \leq b + F\theta \\ & \qquad\qquad\qquad E_{\text{eq}}y = b_{\text{eq}} + F_{\text{eq}}\theta \\ & \qquad\qquad\qquad \theta \in \Theta := \{\theta \in \mathbb{R}^q | \mathrm{CR}_A\theta \leq \mathrm{CR}_b\} \\ & \qquad\qquad\qquad y \in \{0, 1\}^p. \end{aligned} \tag{5.9}$$

This type of problem is referred to as multi-parametric integer linear programming (mp-ILP) problem, and due to its special structure, its solution is significantly less complex than the general mp-MILP problem. This is due to the following two characteristics:

- The solution of problem (5.9) is piecewise constant.
- Given an optimal solution (θ^*, y^*), the critical region CR around θ^* is given as

$$\mathrm{CR} = \left\{ \theta \in \mathbb{R}^q \;\middle|\; \begin{bmatrix} -F \\ -F_{\text{eq}} \\ \mathrm{CR}_A \end{bmatrix} \theta \leq \begin{bmatrix} b - Ey^* \\ b_{\text{eq}} - E_{\text{eq}}y^* \\ \mathrm{CR}_b \end{bmatrix} \right\}, \tag{5.10}$$

where y^* is the optimal solution of problem (5.9) with θ fixed to θ_0, i.e. $\theta = \theta_0$.

Remark 5.4 The inclusion of the equality constraints $E_{eq}y^* = b_{eq} + F_{eq}\theta$ in the critical region description is unique, as in previous chapters this has not been done, as previously we always assumed A_{eq} to have full rank. Thus, it is directly included in the description of the parametric solution, which in return is guaranteed to automatically fulfill the equality constraint. However, as A_{eq} is absent from problem (5.2), the equality constraint needs to be explicitly considered in the critical region description.

Therefore, it is not necessary to solve an mp-LP problem once a candidate solution y^* has been obtained, nor to perform a comparison procedure according to Eq. (5.5). It suffices to identify the optimal solution (θ^*, y^*) over Θ, formulate the corresponding critical region, and explore the remaining part of Θ. The procedure to solve problem (5.9) can then be summarized as follows:

(0) Set $\Omega = \Theta$, $S \leftarrow \emptyset$, where Ω is the remaining parameter space to be explored and S is the solution to problem (5.9).

(1) Treat θ as an optimization variable and solve:

$$\begin{aligned} z(\theta) = \quad & \underset{y,\theta}{\text{Minimize}} \quad c^T y \\ & \text{Subject to} \quad Ey \leq b + F\theta \\ & \qquad E_{eq}y = b_{eq} + F_{eq}\theta \\ & \qquad \theta \in \Omega, \ \ y \in \{0,1\}^p. \end{aligned} \tag{5.11}$$

If problem (5.11) is infeasible, then the algorithm terminates. Otherwise, go to Step 2.

(2) Let y^* be the solution of problem (5.11), then the critical region is defined as

$$\text{CR} = \left\{ \theta \in \Omega \left| \begin{bmatrix} -F \\ -F_{eq} \end{bmatrix} \theta \leq \begin{bmatrix} b - Ey^* \\ b_{eq} - E_{eq}y^* \end{bmatrix} \right. \right\}. \tag{5.12}$$

Note that $\theta \in \Omega$, i.e. θ is restricted to the remaining parameter space to explore. Furthermore, the solution is added to S, i.e. $S \leftarrow (y^*, \text{CR})$. Lastly, the remaining parameter space to be explored Ω is updated as $\Omega = \Omega \backslash \text{CR}$, and the algorithm returns to Step 1.

The key question at this point is how to obtain and describe Ω. As described in Section 1.3,there are two options for this:

- *Constraint reversal*: Each constraint of CR is reversed individually according to[2]

$$R_i = \left\{ \theta \in \Omega \left| \begin{array}{l} -F_i\theta > b_i - E_iy^* \\ A_j\theta \leq b_j - E_jy^*, \ \forall j < i \end{array} \right. \right\}, \quad i = 1, \ldots, m, \tag{5.13}$$

  and thus

$$\begin{aligned} \Omega &= \Omega \backslash \mathrm{CR} \\ &= \{R_1, R_2, \ldots, R_m\}. \end{aligned} \tag{5.14}$$

  As a result, problem (5.11) is solved for all elements of Ω.
- *Integer modelling*: As shown for the mp-QP problems in [1], one of the main drawbacks of constraint reversal approach is the generation of artificial cuts in the parameter space. Thus, consider the case where p optimal solutions have already been identified, i.e. $\mathrm{CR} = \bigcup_{i=1}^{p} \{x | G_i x \leq g_i\}$. In addition, let $G_{i,j}$ and $g_{i,j}$ denote the jth row and element of $G_i \in \mathbb{R}^{t_i \times n}$ and $g_i \in \mathbb{R}^{t_i}$, respectively. Thus, it is possible to describe the difference $\Omega \backslash \mathrm{CR}$ using binary variables:

$$\Omega \backslash \mathrm{CR} \ \Rightarrow \ \left\{ \begin{array}{rl} G_{i,j}^T x + My_i^j & \leq M + g_{i,j} \\ -G_{i,j}^T x + my_i^j & \leq -g_{i,j} \\ t_i y_i - \sum_{j=1}^{t_i} y_i^j & \leq 0 \\ -y_i + \sum_{j=1}^{t_i} y_i^j & \leq t_i - 1 \\ -\sum_{i=1}^{p} y_i & = 0 \end{array} \right. , \tag{5.15}$$

  where M and m are sufficiently large and small, respectively (see Chapter 1.3.3 for details).

2 This procedure can be combined with the removal of redundant constraints in order to reduce the computational effort.

5.4 Chicago to Topeka Featuring a Purchase Decision

The shipping problem introduced in Chapter 2 considered the case where the demand could only be met by transportation from a supplier. In this chapter, we consider the option of purchasing 500 cases of the product for the price of \$100 per case in Topeka. This new problem can be formulated as the following mp-MILP problem:

$$
\begin{aligned}
z(\theta) = \quad & \underset{x,y}{\text{Minimize}} \; c^T \begin{bmatrix} x \\ y \end{bmatrix} \\
& \text{Subject to } x_{\text{Se,Ch}} + x_{\text{Se,To}} \leq 350 \\
& \qquad x_{\text{SD,Ch}} + x_{\text{SD,To}} \leq 600 \\
& \qquad -x_{\text{Se,Ch}} - x_{\text{SD,Ch}} \leq -\theta_1 \\
& \qquad -x_{\text{Se,To}} - x_{\text{SD,To}} - 500y \leq -\theta_2 \\
& \qquad \theta = [\theta_1, \theta_2]^T \in \{\theta \in \mathbb{R}^2 | 0 \leq \theta_i \leq 1000, i = 1, 2\} \\
& \qquad x = [x_{\text{Se,Ch}}, x_{\text{Se,To}}, x_{\text{SD,Ch}}, x_{\text{SD,To}}]^T \geq 0. \; y \in \{0, 1\}
\end{aligned}
\tag{5.16}
$$

where

$$
c^T \begin{bmatrix} x \\ y \end{bmatrix} = \begin{bmatrix} 178 & 187 & 187 & 151 & 50000 \end{bmatrix} \begin{bmatrix} x_{\text{Se,Ch}} \\ x_{\text{Se,To}} \\ x_{\text{SD,Ch}} \\ x_{\text{SD,To}} \\ y \end{bmatrix}. \tag{5.17}
$$

The solution of this problem is given by seven critical regions (see Figure 5.5).

5.4.1 Interpretation of the Results

In Table 5.1 and Figure 5.5, it is shown that the solution to problem (5.16) is given by seven critical regions. On the contrary to previous chapters, here the discussion is based not on each critical region, but on the difference between the regions where $y = 0$ and those where $y = 1$.

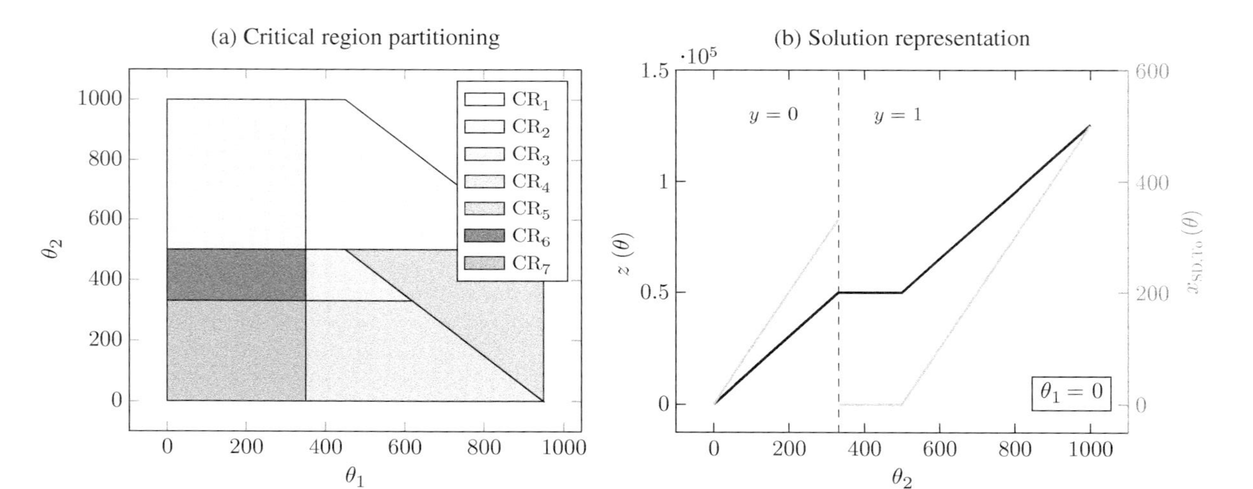

Figure 5.5 The solution of problem (5.16). In (a) the partitioning of the parameter space into seven critical regions is shown, and in (b) the optimizer $x_{SD,To}(\theta)$, binary variable y, and optimal objective function $z(\theta)$ are shown as a function of the parameter θ_2 along the line of $\theta_1 = 0$, i.e. no demand from Chicago. Note that the dashed line simply highlights the switch between $y = 0$ and $y = 1$, and that the left y-axis is associated with $z(\theta)$ (black line) and the right y-axis is associated with $x(\theta)$ (gray line).

Table 5.1 The solution of problem (5.16).

$$\text{CR}_1 \begin{cases} \text{CR} = \{\theta \mid \theta_1 \geq 350, 500 \leq \theta_2 \leq 1000, \theta_1 + \theta_2 \leq 1450\} \\ x(\theta) = [350, 0, \theta_1 - 350, \theta_2 - 500]^T \\ y = 1 \\ z(\theta) = 187\theta_1 + 151\theta_2 - 28650 \end{cases}$$

$$\text{CR}_2 \begin{cases} \text{CR} = \{\theta \mid 0 \leq \theta_1 \leq 350, 500 \leq \theta_2 \leq 1000\} \\ x(\theta) = [\theta_1, 0, 0, \theta_2 - 500]^T \\ y = 1 \\ z(\theta) = 178\theta_1 + 151\theta_2 - 25500 \end{cases}$$

$$\text{CR}_3 \begin{cases} \text{CR} = \{\theta \mid \theta_1 \geq 350, 331.1 \leq \theta_2 \leq 500, \theta_1 + \theta_2 \leq 950\} \\ x(\theta) = [350, 0, \theta_1 - 350, 0]^T \\ y = 1 \\ z(\theta) = 187\theta_1 + 46350 \end{cases}$$

$$\text{CR}_4 \begin{cases} \text{CR} = \{\theta \mid \theta_1 \geq 350\ 0 \leq \theta_2 \leq 331.1, \theta_1 + \theta_2 \leq 950\} \\ x(\theta) = [350, 0, \theta_1 - 350, \theta_2]^T \\ y = 0 \\ z(\theta) = 187\theta_1 + 151\theta_2 - 31\,500 \end{cases}$$

$$\text{CR}_5 \begin{cases} \text{CR} = \{\theta \mid 350 \leq \theta_1 \leq 950, 0 \leq \theta_2 \leq 500, \theta_1 + \theta_2 \geq 950\} \\ x(\theta) = [350, 0, \theta_1 - 350, 0]^T \\ y = 1 \\ z(\theta) = 187\theta_1 + 46850 \end{cases}$$

$$\text{CR}_6 \begin{cases} \text{CR} = \{\theta \mid 0 \leq \theta_1 \leq 350, 331.1 \leq \theta_2 \leq 500\} \\ x(\theta) = [\theta_1, 0, 0, 0]^T \\ y = 1 \\ z(\theta) = 178\theta_1 + 50\,000 \end{cases}$$

$$\text{CR}_7 \begin{cases} \text{CR} = \{\theta \mid 0 \leq \theta_1 \leq 350, 0 \leq \theta_2 \leq 331.1\} \\ x(\theta) = [\theta_1, 0, 0, \theta_2]^T \\ y = 0 \\ z(\theta) = 178\theta_1 + 151\theta_2 \end{cases}$$

Regions where $y = 0$: For regions CR_4 and CR_7, the binary variable $y = 0$. This in turn means that the purchasing option is not taken, and that the known schedule from Chapter 2 is pursued. As can be seen in Figure 5.5b, the reasons for this is that it is cheaper to ship without the option than to pay the one-time \$50 000 fee for purchasing 500 cases in Topeka, which in return is driven by the low demand (i.e. parameter realizations) as well as the cost of the shipping itself. Note that as soon as the cost of shipping surpasses \$50 000 overall, the purchasing option is taken. In particular, when the demand surpasses $\frac{50\,000}{151} = 331.1$ cases, then the purchase is made, where 151 is the cost of shipping from San Diego to Topeka, which is cheaper than from Seattle to Topeka (187).

Regions where $y = 1$: For regions CR_1, CR_2, CR_3, CR_5, and CR_6, the binary variable $y = 1$. Thus, the 500 cases are purchased in Topeka, and the demand in Topeka θ_2 is reduced by the equivalent amount. The resulting new problem is solved according to the principles from Chapter 2.

5.5 Literature Review

Despite the high level of complexity from a theoretical and implementation point of view, (mixed)-integer linear programming featuring a single parameter was considered as early as 1966, and gained substantial interest in the 1970s and 1980s [2–6]. The main focus at the time was thereby to show that it is conceptually possible to devise an algorithm that rigorously solves the parametric problem. These developments were captured in detail an excellent survey by Arthur Geoffrion and Robert Nauss [7] and an annotated bibliography by Harvey Greenberg [8].

However, despite the ability to solve mp-LP problems systematically, which was introduced in 1972 by Gal and Nedoma [9], it was not until 1997 that the first solution strategy for mp-MILP problems was proposed by Acevedo and Pistikopoulos [10], which was based on the branch-and-bound approach. This approach was followed by the seminal paper by Dua and Pistikopoulos [11], which described the global-optimization based approach still in use today.[3] Another take on the branch and bound method was presented in [13], where the mp-LP problem is solved only at the leaf node of the optimal solution, and the comparison procedure is started from there. These developments led to the consideration of left-hand side uncertainty, i.e.

$$A(\theta)x + E(\theta)y \leq b + F\theta, \tag{5.18}$$

where [14] extended the approach by Jia and Ierapetritou [13], while Wittmann-Hohlbein and Pistikopoulos presented a two-stage approach combining the global optimization approach in [11] with appropriate McCormick relaxation techniques [15–17]. In addition, Mitsos and Barton have revisited the single parameter case in 2009 and proposed two solution methods based on a branch-and-bound approach and the optimality range of a qualitatively invariant solution [18].

On the contrary to mp-MILP problems, the consideration of the multi-parametric version of integer linear programming problems went widely unnoticed. The notable exception is the work of Alejandro Crema, who considered several special cases of mp-ILP problems [19–23] before combining these efforts into a unified theory [24].

References

1 Tøndel, P., Johansen, T.A., and Bemporad, A. (2003) An algorithm for multi-parametric quadratic programming and explicit MPC solutions. *Automatica*, 39 (3), 489–497, doi: 10.1016/S0005-1098(02)00250-9. URL http://www.sciencedirect.com/science/article/pii/S0005109802002509.

2 Marsten, R.E. and Morin, T.L. (1977) Parametric integer programming: the right-hand-side case, in *Studies in integer programming, Annals of discrete mathematics*, vol. 1 (eds P.L. Hammer, E.L. Johnson, B.H. Korte, and G.L. Nemhauser), Elsevier, pp. 375–390, doi: 10.1016/S0167-5060(08)70745-3.

3 Note that already one year earlier Dua and Pistikopoulos published the global-optimization based approach for the case of general multi-parametric mixed-integer nonlinear programming problems [12].

URL http://www.sciencedirect.com/science/article/pii/S0167506008707453.

3 Klein, D. and Holm, S. (1979) Integer programming post-optimal analysis with cutting planes. *Management Science*, 25 (1), 64–72, doi: 10.1287/mnsc.25.1.64. URL http://dx.doi.org/10.1287/mnsc.25.1.64.

4 Bailey, M.G. and Gillett, B.E. (1980) Parametric integer programming analysis: a contraction approach. *The Journal of the Operational Research Society*, 31 (3), 257–262. URL http://www.jstor.org/stable/2581082.

5 Rountree, S.L. and Gillett, B.E. (1982) Parametric integer linear programming: a synthesis of branch and bound with cutting planes. *European Journal of Operational Research*, 10 (2), 183–189, doi: 10.1016/0377-2217(82)90158-8. URL http://www.sciencedirect.com/science/article/pii/0377221782901588.

6 Holm, S. and Klein, D. (1984) *Three methods for postoptimal analysis in integer linear programming: sensitivity, stability and parametric analysis*, Springer-Verlag, Berlin, Heidelberg, pp. 97–109, doi: 10.1007/BFb0121213. URL http://dx.doi.org/10.1007/BFb0121213.

7 Geoffrion, A.M. and Nauss, R. (1977) Parametric and postoptimality analysis in integer linear programming. *Management Science*, 23 (5), 453–466, doi: 10.1287/mnsc.23.5.453.

8 Greenberg, H.J. (1998) An annotated bibliography for post-solution analysis in mixed integer programming and combinatorial optimization, in (eds. D.L. Woodruff)*Advances in computational and stochastic optimization, logic programming, and heuristic search, Interfaces in computer science and operations research*, Springer US, Boston, MA, pp. 97–147, doi: 10.1007/978-1-4757-2807-1_ 4. URL http://dx.doi.org/10.1007/978-1-4757-2807-1. 4.

9 Gal, T. and Nedoma, J. (1972) Multiparametric linear programming. *Management Science*, 18 (7), 406–422, doi: 10.1287/mnsc.18.7.406.

10 Acevedo, J. and Pistikopoulos, E.N. (1997) A multiparametric programming approach for linear process engineering problems under uncertainty. *Industrial and Engineering Chemistry Research*, 36 (3), 717–728, doi: 10.1021/ie960451l.

11 Dua, V. and Pistikopoulos, E.N. (2000) An algorithm for the solution of multiparametric mixed integer linear programming problems. *Annals of Operations Research*, 99 (1–4), 123–139, doi: 10.1023/A:1019241000636. URL http://dx.doi.org/10.1023/A%3A1019241000636.

12 Dua, V. and Pistikopoulos, E.N. (1999) Algorithms for the solution of multiparametric mixed-integer nonlinear optimization problems. *Industrial and Engineering Chemistry Research*, 38 (10), 3976–3987, doi: 10.1021/ie980792u.

13 Jia, Z. and Ierapetritou, M.G. (2006) Uncertainty analysis on the righthand side for MILP problems. *AIChE Journal*, 52 (7), 2486–2495, doi: 10.1002/aic.10842. URL http://dx.doi.org/10.1002/aic.10842.

14 Li, Z. and Ierapetritou, M.G. (2007) A new methodology for the general multiparametric mixed-integer linear programming (MILP) problems. *Industrial and Engineering Chemistry Research*, 46 (15), 5141–5151, doi: 10.1021/ie070148s.

15 Wittmann-Hohlbein, M. and Pistikopoulos, E.N. (2012) A two-stage method for the approximate solution of general multiparametric mixed-integer linear programming problems. *Industrial and Engineering Chemistry Research*, 51 (23), 8095–8107, doi: 10.1021/ie201408p.

16 Wittmann-Hohlbein, M. and Pistikopoulos, E.N. (2013) On the global solution of multi-parametric mixed integer linear programming problems. *Journal of Global Optimization*, 57 (1), 51–73, doi: 10.1007/s10898-012-9895-2. URL http://dx.doi.org/10.1007/s10898-012-9895-2.

17 Wittmann-Hohlbein, M. and Pistikopoulos, E.N. (2014) Approximate solution of mp-MILP problems using piecewise affine relaxation of bilinear terms. *Computers and Chemical Engineering*, 61, 136–155, doi: 10.1016/j.compchemeng.2013.10.009. URL http://www.sciencedirect.com/science/article/pii/S0098135413003311.

18 Mitsos, A. and Barton, P.I. (2009) Parametric mixed-integer 0–1 linear programming: the general case for a single parameter. *European Journal of Operational Research*, 194 (3), 663–686, doi: 10.1016/j.ejor.2008.01.007. URL http://www.sciencedirect.com/science/article/pii/S0377221708001331.

19 Crema, A. (1997) A contraction algorithm for the multi-parametric integer linear programming problem. *European Journal of Operational Research*, 101 (1), 130–139, doi: 10.1016/0377-2217(95)00369-X. URL http://www.sciencedirect.com/science/article/pii/037722179500369X.

20 Crema, A. (1999) An algorithm to perform a complete right-hand-side parametrical analysis for a 0–1-integer linear programming problem. *European Journal of Operational Research*, 114 (3), 569–579, doi: 10.1016/S0377-2217(98)00132-5. URL http://www.sciencedirect.com/science/article/pii/S0377221798001325.

21 Crema, A. (2000) An algorithm for the multiparametric 0–1-integer linear programming problem relative to the constraint matrix. *Operations Research Letters*, 27 (1), 13–19, doi: 10.1016/S0167-6377(00)00034-1. URL http://www.sciencedirect.com/science/article/pii/S0167637700000341.

22 Crema, A. (2000) An algorithm for the multiparametric 0–1-integer linear programming problem relative to the objective function. *European Journal of Operational Research*, 125 (1), 18–24, doi: 10.1016/S0377-2217(99)00193-9. URL http://www.sciencedirect.com/science/article/pii/S0377221799001939.

23 Crema, A. (2002) An algorithm to perform a complete parametric analysis relative to the constraint matrix for a 0–1-integer linear program. *European Journal of Operational Research*, 138 (3), 484–494, doi: 10.1016/S0377-2217(01)00162-X. URL http://www.sciencedirect.com/science/article/pii/S037722170100162X.

24 Crema, A. (2002) The multiparametric 0–1-integer linear programming problem: a unified approach. *European Journal of Operational Research*, 139 (3), 511–520, doi: 10.1016/S0377-2217(01)00163-1. URL http://www.sciencedirect.com/science/article/pii/S0377221701001631.

6

Multi-parametric Mixed-integer Quadratic Programming

Consider the following mixed-integer quadratic programming (MIQP) problem:

$$
\begin{aligned}
z = \underset{x,y}{\text{Minimize}} \quad & (Q\omega + c)^T \omega \\
\text{Subject to} \quad & Ax + Ey \leq b \\
& A_{\text{eq}} x + E_{\text{eq}} y = b_{\text{eq}} \\
& x \in \mathbb{R}^n, \ y \in \{0,1\}^p, \ \omega = [x^T \ y^T]^T,
\end{aligned}
\tag{6.1}
$$

where $Q \in \mathbb{R}^{(n+p)\times(n+p)}$ is symmetric positive definite and $c \in \mathbb{R}^{(n+p)}$, $A \in \mathbb{R}^{m\times n}$, $E \in \mathbb{R}^{m\times p}$, $b \in \mathbb{R}^m$, $A_{\text{eq}} \in \mathbb{R}^{s\times n}$, $E_{\text{eq}}\mathbb{R}^{s\times p}$, $b_{\text{eq}} \in \mathbb{R}^s$ and $Ax \leq (b - E\bar{y})$ is a compact polytope for all feasible $\bar{y} \in \{0,1\}^p$. Problems of type (6.1) occur in many instances where a system featuring discrete elements is considered based on a quadratic penalty function. One of the most common application is thereby the optimal control of hybrid systems, where analogously to Chapter 3 the classical consideration of a quadratic error gives rise to an MIQP problem. However, especially for dynamic problems such as optimal control problems, values such as states, property values, and tuning parameters might change over time and have great impact on the solution. In order to consider such scenarios directly, the uncertainty can be considered explicitly by formulating a multi-parametric mixed-integer quadratic

Multi-parametric Optimization and Control, First Edition.
Efstratios N. Pistikopoulos, Nikolaos A. Diangelakis, and Richard Oberdieck.

programming (mp-MIQP) problem:

$$\begin{aligned} z(\theta) = \underset{x,y}{\text{Minimize}} \quad & (Q\omega + H\theta + c)^T \omega \\ \text{Subject to} \quad & Ax + Ey \leq b + F\theta \\ & A_{\text{eq}}x + E_{\text{eq}}y = b_{\text{eq}} + F_{\text{eq}}\theta \\ & \theta \in \Theta := \{\theta \in \mathbb{R}^q | \text{CR}_A\theta \leq \text{CR}_b\} \\ & x \in \mathbb{R}^n, \; y \in \{0,1\}^p, \; \omega = [x^T \; y^T]^T, \end{aligned} \tag{6.2}$$

where $H \in \mathbb{R}^{(n+p)\times q}$, $F \in \mathbb{R}^{m\times q}$, $F_{\text{eq}} \in \mathbb{R}^{s\times q}$, $\text{CR}_A \in \mathbb{R}^{r\times q}$, $\text{CR}_b \in \mathbb{R}^r$ and Θ is a compact polytope. Note that it is assumed that there do not exist any equality constraints featuring only the parameters, as this would render the parameter space lower-dimensional. The key difference between problems (6.1) and (6.2) is thereby the presence of the bounded uncertain parameters θ in the objective function and the constraints, which are not considered as optimization variables. As a result, the solution x and the optimal objective function z of problem (6.2) are obtained as a function of θ, i.e. $x(\theta)$ and $z(\theta)$, respectively. Note that it is possible to add a scalar d to the objective function of an MIQP problem without influencing the optimal solution. Similarly, it is possible to add an arbitrary scaling function $f(\theta)$ to an mp-MIQP problem without influencing the optimal solution.

Remark 6.1 Although the term "mixed-integer" indicates that general integer variables are considered, i.e. $d = \{n_{d_0}, n_{d_0} + 1, \ldots, n_{d_1}\}^p$, in this chapter we only consider binary variables. Note however that it is possible to model any integer variable d as a set of $(d_0 + d_1) + 1$ binary variables, $y_i \in \{0,1\}$, via

$$\sum_{i=n_{d_0}}^{n_{d_1}} iy_i = d \tag{6.3a}$$

$$\sum_{i=n_{d_0}}^{n_{d_1}} y_i = 1. \tag{6.3b}$$

6.1 Solution Properties

6.1.1 From mp-QP to mp-MIQP Problems

Conceptually, the presence of binary variables can be understood as a set of discrete choices, each of which can be selected by fixing the binary variables to the corresponding value, resulting in an mp-QP problem of type (3.2). Thus, problem (6.2) can be regarded as 2^p different mp-QP problems. As a result, it is clear that the solution properties of mp-MIQP problems will consist of the solution properties of mp-QP problems combined with the changes incurred by considering more than one solution in the same parameter space. This concept is schematically depicted in Figure 6.1. The main difference to the equivalent discussion for mp-MILP problems of type (5.2) is the quadratic nature of the objective function, which results in a non-polytopic partitioning of the parameter space and thus requires a more careful consideration.

6.1.2 The Properties

Theorem 6.1 *Consider the optimal solution of problem (6.2) with Q symmetric positive definite. Then the set of feasible parameters $\Theta_f \subseteq \Theta$ is a possibly non-convex union of quadratically constrained critical regions of the form:*

$$\mathrm{CR}_i = \{\theta | \theta^T G_{i,j} \theta + h_{i,j}^T \theta \leq w_{i,j}, j = 1, \ldots, t_i\}, \tag{6.4}$$

where t_i is the number of constraints that describe CR_i, and the optimizer $x(\theta) : \Theta_f \mapsto \mathbb{R}^n$ and the optimal objective function $z(\theta) : \Theta_f \mapsto \mathbb{R}$ are piecewise affine and quadratic, respectively.

Proof: The piecewise affine and quadratic nature of $x(\theta)$ and $z(\theta)$, respectively, follows directly from Theorem 5.1. In turn, the nature of the critical regions is a result of the necessity to compare the solution of several solutions of mp-QP problems, which feature piecewise quadratic objective functions. As a result,

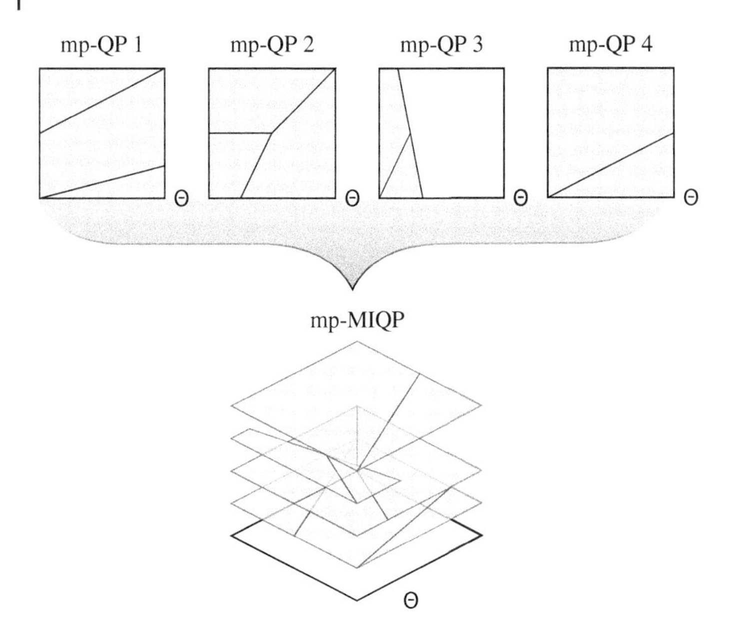

Figure 6.1 The interpretation of mp-MIQP problems as a combination of several mp-QP problems.

the comparison of such solutions yields quadratic boundaries, which leads to the description of critical regions according to Eq. (6.4). □

Lemma 6.1 ***(Quadratic Boundaries)*** *Quadratic boundaries arise from the comparison of quadratic objective functions associated with the solution of mp-QP problems for different feasible combinations of binary variables.*

Proof: This follows directly from Theorem 6.1. □

6.2 Comparing the Solutions from Different mp-QP Problems

Remark 6.2 Due to the similarities between the solution procedures for mp-MILP and mp-MIQP problems, they are described together in Chapter 7.

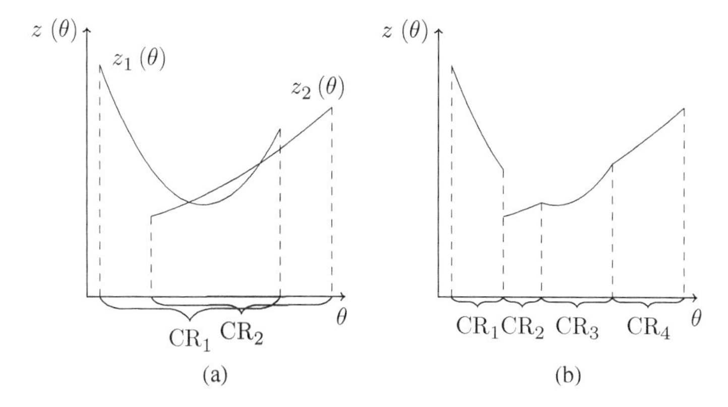

Figure 6.2 The schematic representation of the example of a comparison procedure for an mp-MIQP problem. In (a), the (quadratic) optimal objective function values of a critical region from two mp-QP problems are shown as a function of the parameter. As they overlap, (b) shows the resulting optimal objective function value resulting from the comparison of the two optimal objective functions for the critical regions, over which they are defined. Note that the choice of which solution is optimal is purely driven by the need to obtain a lower objective function value.

Similarly to the mp-MILP problems, the key operation for mp-MIQP problems is the comparison of several parametric solutions. In particular, the solution of an mp-QP problems of type (3.2) is given by its feasible parameter space Θ_f partitioned into critical regions CR_i, associated solutions $x_i(\theta)$, and optimal objective functions $z_i(\theta)$. An example of a comparison procedure is schematically shown in Figure 6.2.

Thus, in order to compare the solutions from two mp-QP problems,[1] the following operations need to be performed.

Remark 6.3 In order to distinguish the two solutions, all elements associated with the solution of the second mp-QP problem are denoted with a bar, i.e. $\bar{\cdot}$.

---

1 Once the tools to compare the solutions from two mp-QP problems have been developed, it is straightforward to extend them to the case of n mp-QP problems.

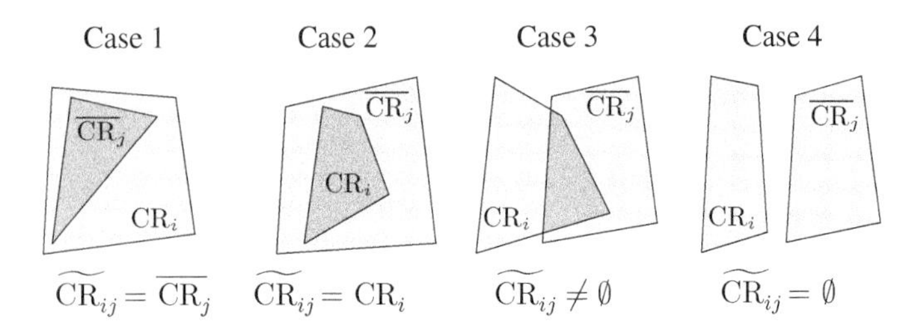

Figure 6.3 The detection of the intersection between two critical regions, $\widetilde{CR}_{ij} = \overline{CR}_j \cap CR_i$.

6.2.1 Identification of overlapping critical regions

In order to compare two optimal objective functions, $z_i(\theta)$ and $\overline{z}_j(\theta)$, they need to be valid in the same region of the parameter space. Thus, every combination of CR_i and $\overline{CR}_j$, needs to be classified into one of the following cases (see Figure 6.3):

- *Case* 1: $\widetilde{CR}_{ij} = \overline{CR}_j \cap CR_i = \overline{CR}_j$.
- *Case* 2: $\widetilde{CR}_{ij} = \overline{CR}_j \cap CR_i = CR_i$.
- *Case* 3: $\widetilde{CR}_{ij} = \overline{CR}_j \cap CR_i \neq \emptyset$.
- *Case* 4: $\widetilde{CR}_{ij} = \overline{CR}_j \cap CR_i = \emptyset$.

If $\widetilde{CR}_{ij} \neq \emptyset$ (i.e. Case 1-3), then the two solutions need to be compared against each other.

6.2.2 Performing the Comparison

Once an overlap $\widetilde{CR}_{ij}$ has been identified, the comparison between the two solutions can be expressed as

$$z(\theta) = \min \ \{\overline{z}_j(\theta), z_i(\theta)\}, \tag{6.5}$$

which directly leads to the following difference equation:

$$\Delta z(\theta) = \overline{z}_j(\theta) - z_i(\theta) \overset{!}{=} 0. \tag{6.6}$$

As a result, the solution of Eq. (5.5) is a partitioning of $\widetilde{CR}_{ij}$ according to (see Figure 6.4b):

$$\widetilde{CR}^1_{ij} = \widetilde{CR}_{ij} \cap \Delta z(\theta) \leq 0 \qquad \Rightarrow \overline{z}_j(\theta) \text{ optimal} \tag{6.7a}$$

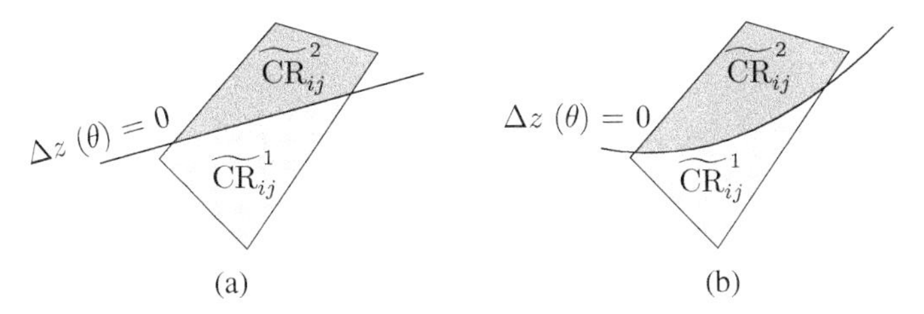

Figure 6.4 The difference in the partitioning of a critical region due to comparison for (a) mp-MILP problems based on Eq. (5.6) and (b) mp-MIQP problems based on Eq. (6.5). Due to the quadratic nature of the objective function, quadratically constrained critical regions may result.

$$\widetilde{CR}_{ij}^2 = \widetilde{CR}_{ij} \cap \Delta z(\theta) \geq 0 \qquad \Rightarrow z_i(\theta) \text{ optimal.} \tag{6.7b}$$

Note that since $\overline{z}_j(\theta)$ and $z_i(\theta)$ may be quadratic, $\widetilde{CR}_{ij}^1$ and $\widetilde{CR}_{ij}^2$ potentially are quadratically constrained critical regions. This is a fundamental difference to the mp-MILP case, and is shown schematically in Figure 6.4. As a result, while in the mp-MILP case a simple constraint reversal is used to ensure that the entire parameter space is explored, this is not applicable to mp-MIQP cases. This has motivated the construction of so-called envelopes of solutions, which are discussed in the next section.

6.3 Envelope of Solutions

As shown in Figure 6.4b, the direct use of $\Delta z(\theta)$ in order to partition the parameter space may lead to quadratically constrained critical regions. This is a fundamental increase in complexity, both from a solution property and a solution procedure point of view. In particular, all the useful and computationally attractive properties of polytopes are no longer applicable.

This issue has led to the consideration of so-called envelopes of solutions:

Definition 6.1 (Envelope of Solutions).In order to avoid the non-convex critical regions described by Lemma 6.1, an envelope of solutions is created where more than one solution is associated

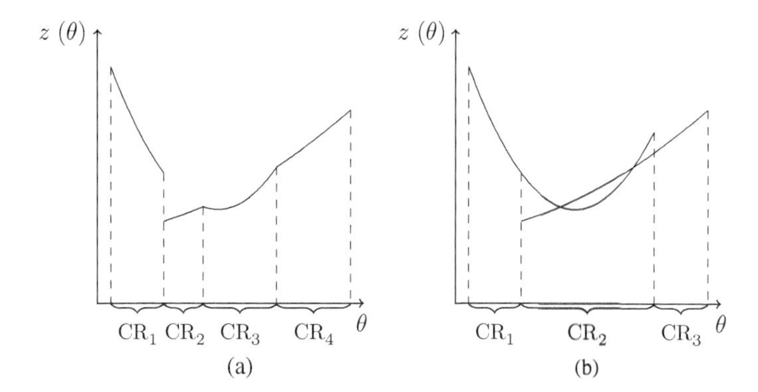

Figure 6.5 The schematic representation of the envelopes of solutions. In (a), Figure 6.2b is shown as an example of the exact solution of an mp-MIQP problem, while (b) shows the storage of both solutions in the overlapping part of the parameter space. Note that the number of critical region changes as well as the fact that in order to evaluate the optimal solution for the envelope of solutions portrayed in (b), a pointwise comparison procedure is required.

with one critical region. The envelope is guaranteed to contain the optimal solution, and a point-wise comparison procedure among the envelope of solutions is performed online.

Thus, an envelope of solutions effectively stores more than one parametric solution $(x(\theta), y, z(\theta))$ in a single critical region CR. This circumvents the comparison procedure and thus retains the polytopic nature of the critical regions. In order to illustrate the concept, Figure 6.5 shows the example comparison of Figure 6.2 with and without envelopes of solutions.

6.4 Chicago to Topeka Featuring Quadratic Cost and A Purchase Decision

The shipping problem introduced in Chapter 3 considered the case where the demand could only be met by transportation from a supplier. In this chapter, we consider the option of purchasing 500 cases of the product for the total price of $7.6 million in Topeka.

Thus, the overall cost function $J(x)$ is given as

$$\begin{aligned} J(\omega) &= \omega^T \operatorname{diag}(c_{\text{dist}} \cdot d_{i,j})\omega + \begin{bmatrix} c_{\text{load}} \\ 7.6 \times 10^6 \end{bmatrix}^T \omega \\ &= \omega^T \begin{bmatrix} 153 & 0 & 0 & 0 & 0 \\ 0 & 162 & 0 & 0 & 0 \\ 0 & 0 & 162 & 0 & 0 \\ 0 & 0 & 0 & 126 & 0 \\ 0 & 0 & 0 & 0 & 0 \end{bmatrix} \omega + \begin{bmatrix} 25 \\ 25 \\ 25 \\ 25 \\ 7.6 \times 10^6 \end{bmatrix} \omega, \end{aligned} \tag{6.8}$$

where $\omega = [x^T \ y^T]^T$, $x = [x_{\text{Se,Ch}}, x_{\text{Se,To}}, x_{\text{SD,Ch}}, x_{\text{SD,To}}]^T$, $y \in \{0, 1\}$, and $\operatorname{diag}(a)$ denotes a diagonal matrix where the weights are specified by the vector a.

Remark 6.4 It would have been equally possible to express the cost of the binary variable in the linear cost vector, as $y = y^2$ for $y \in \{0, 1\}$. The quadratic term was chosen in order to obtain a positive definite quadratic matrix.

This new problem can be formulated as the following mp-MIQP problem:

$$\begin{aligned} z(\theta) = \ & \underset{x,y}{\text{Minimize}} J(\omega) \\ & \text{Subject to } x_{\text{Se,Ch}} + x_{\text{Se,To}} \le 350 \\ & x_{\text{SD,Ch}} + x_{\text{SD,To}} \le 600 \\ & -x_{\text{Se,Ch}} - x_{\text{SD,Ch}} \le -\theta_1 \\ & -x_{\text{Se,To}} - x_{\text{SD,To}} - 500y \le -\theta_2 \\ & \theta = [\theta_1, \theta_2]^T \in \{\theta \in \mathbb{R}^2 | 0 \le \theta_i \le 1000, i = 1, 2\} \\ & x = [x_{\text{Se,Ch}}, x_{\text{Se,To}}, x_{\text{SD,Ch}}, x_{\text{SD,To}}]^T \ge 0, \\ & y \in \{0, 1\}, \ \omega = [x^T \ y^T]^T. \end{aligned} \tag{6.9}$$

The solution of this problem is given by eight critical regions, shown in Figure 6.6 and partially reported in Table 6.1.

6.4.1 Interpretation of the Results

In Figure 6.6, it is shown that the solution to problem (6.9) is given by eight critical regions, six of which are shown in Table 6.1. Note

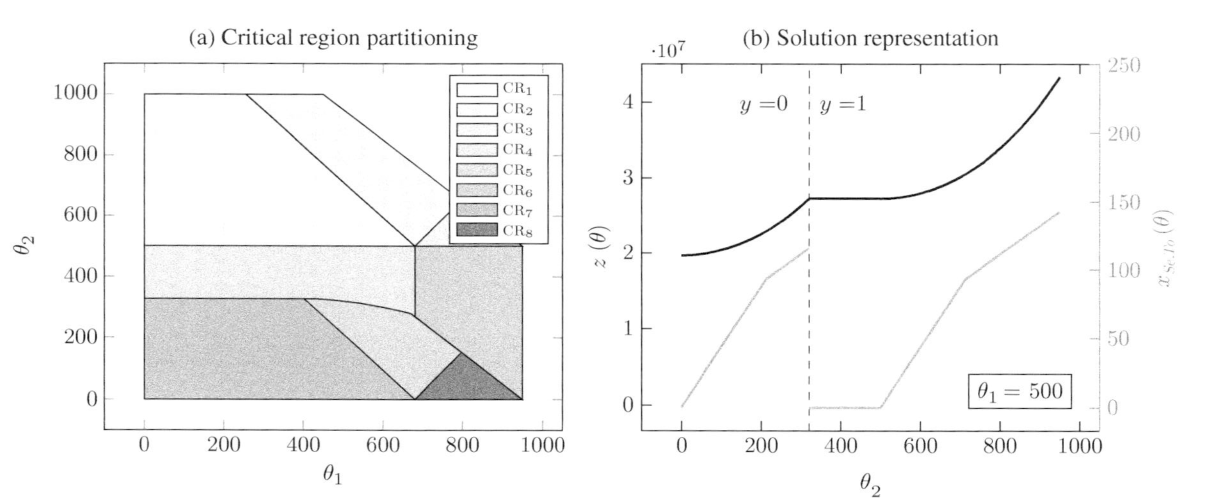

Figure 6.6 The solution of problem (6.9). In (a) the partitioning of the parameter space into eight critical regions is shown, and in (b) the optimizer $x(\theta)$, binary variable y and optimal objective function $z(\theta)$ are shown as a function of the parameter θ_2 along the line of $\theta_1 = 500$, i.e. a fixed demand of 500 cases from Chicago. Note that the dashed line simply highlights the switch between $y = 0$ and $y = 1$, and that the left y-axis is associated with $z(\theta)$ (black line) and the right y-axis is associated with $x_{\text{Se,To}}(\theta)$ (gray line).

Table 6.1 The partial solution of problem (6.9). Note that the critical regions CR_6 and CR_8 are omitted for conciseness as they similar to CR_1, i.e. they are polytopes. While CR_1 is reported as a representative member of a polytopic critical region, CR_4 and CR_5 feature quadratic constraints.

CR_1	$\mathrm{CR} = \left\{\theta \,\middle|\, \begin{array}{c} \theta_1 + 0.85\theta_2 \leq 1105.9, \theta_1 \geq 0, \\ 500 \leq \theta_2 \leq 1000 \end{array}\right\}$ $x(\theta) = \begin{bmatrix} 0.51 & 0 \\ 0 & 0.44 \\ 0.49 & 0 \\ 0 & 0.56 \end{bmatrix}\theta + \begin{bmatrix} 0 \\ -218.75 \\ 0 \\ -218.75 \end{bmatrix}$ $y = 1$
CR_4	$\mathrm{CR} = \left\{\theta \,\middle|\, \begin{array}{c} 39.79\theta_1^2 + 67.7\theta_1\theta_2 + 99.67\theta_2^2 - 54\,161\theta_1 \cdots \\ \cdots - 46\,049.63\theta_2 \geq -1.08 \cdot 10^7, \forall\theta_1 \leq 402.14 \\ \theta_2 \geq 327.64, \forall\theta_1 \geq 402.14 \\ 0 \leq \theta_1 \leq 680.56, \theta_2 \leq 500 \end{array}\right\}$ $x(\theta) = \begin{bmatrix} 0.51 & 0 \\ 0 & 0 \\ 0.49 & 0 \\ 0 & 0 \end{bmatrix}\theta$ $y = 1$
CR_2	$\mathrm{CR} = \left\{\theta \,\middle|\, \begin{array}{c} \theta_1 + \theta_2 \leq 1450, \theta_1 - 0.78\theta_2 \leq 291.67, \\ \theta_1 - 0.85\theta_2 \geq 1105.9, \theta_2 \leq 1000 \end{array}\right\}$ $x(\theta) = \begin{bmatrix} 0.27 & -0.21 \\ -0.27 & 0.21 \\ 0.73 & 0.21 \\ 0.27 & 0.79 \end{bmatrix}\theta + \begin{bmatrix} 271.64 \\ 78.36 \\ -271.64 \\ -578.36 \end{bmatrix}$ $y = 1$
CR_5	$\mathrm{CR} = \left\{\theta \,\middle|\, \begin{array}{c} 39.79\theta_1^2 + 67.7\theta_1\theta_2 + 99.67\theta_2^2 - 54\,161\theta_1 \cdots \\ \cdots - 46\,049.63\theta_2 \leq -1.08 \cdot 10^7, \forall\theta_1 \leq 670.69 \\ \theta_1 + \theta_2 \leq 950, \forall\theta_1 \geq 670.69, \\ \theta_1 + 0.85\theta_2 \geq 680.56, \\ \theta_1 - 0.78\theta_2 \leq 680.56, \theta_1 \geq 401.83 \end{array}\right\}$ $x(\theta) = \begin{bmatrix} 0.27 & -0.21 \\ -0.27 & 0.21 \\ 0.73 & 0.21 \\ 0.27 & 0.79 \end{bmatrix}\theta + \begin{bmatrix} 167.16 \\ 182.84 \\ -167.16 \\ -182.84 \end{bmatrix}$ $y = 0$

Table 6.1 (Continued)

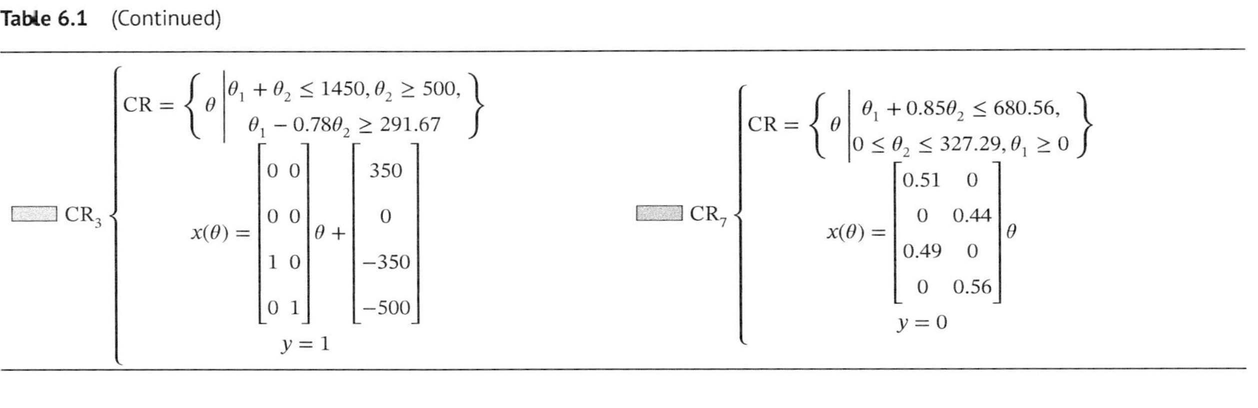

CR_3	$\begin{cases} \text{CR} = \left\{ \theta \,\middle\vert\, \begin{array}{c} \theta_1 + \theta_2 \leq 1450, \theta_2 \geq 500, \\ \theta_1 - 0.78\theta_2 \geq 291.67 \end{array} \right\} \\ x(\theta) = \begin{bmatrix} 0 & 0 \\ 0 & 0 \\ 1 & 0 \\ 0 & 1 \end{bmatrix} \theta + \begin{bmatrix} 350 \\ 0 \\ -350 \\ -500 \end{bmatrix} \\ y = 1 \end{cases}$
CR_7	$\begin{cases} \text{CR} = \left\{ \theta \,\middle\vert\, \begin{array}{c} \theta_1 + 0.85\theta_2 \leq 680.56, \\ 0 \leq \theta_2 \leq 327.29, \theta_1 \geq 0 \end{array} \right\} \\ x(\theta) = \begin{bmatrix} 0.51 & 0 \\ 0 & 0.44 \\ 0.49 & 0 \\ 0 & 0.56 \end{bmatrix} \theta \\ y = 0 \end{cases}$

that on the contrary to previous chapters, not the entire solution is reported for conciseness reasons, and the focus is put on the quadratically constrained critical regions. The other critical regions are structurally equivalent to the mp-MILP counterparts. The discussion is based not on each critical region, but on the transition from $y = 0$ to $y = 1$.

Regions where $y = 0$: For regions CR_5, CR_7, and CR_8, the binary variable $y = 0$. This in turn means that the purchasing option is not taken, and that the known schedule from Chapter 3 is pursued. As can be seen in Figure 6.6b, the reason for this is that it is cheaper to ship than to pay the one-time fee for purchasing 500 cases in Topeka, which in return is driven by the low demand (i.e. parameter realizations) as well as the cost of the shipping itself. Note that as soon as the cost of shipping surpasses the fee cost, the purchasing option is taken. On the contrary to the mp-MILP case, where the change from $y = 0$ to $y = 1$ was purely driven by the shipping cost from San Diego to Topeka, here the quadratic nature of the objective function yields a more diversified shipping route and as such, the transition line depends on the overall cost incurred (i.e. quadratically characterized critical region partitioning between CR_4 and CR_5).

Regions where $y = 1$: For the remaining regions, the binary variable $y = 1$. Thus, the 500 cases are purchased in Topeka, and the demand in Topeka θ_2 is reduced by the equivalent amount. The resulting new problem is solved according to the principles from Chapter 3. Note that Figure 6.6b highlights the quadratic nature of the objective function, as well as the plateau reached after the purchase decision has been taken, as the demand in Topeka is completely satisfied.

6.5 Literature Review

Despite its complexity, the idea of solving MIQP problems (multi)-parametrically has been considered very early on, with a major peak of interest in the 1980s [1–5]. However, it was not until it was possible to solve mp-QP problems exactly that in 2002 Dua *et al.* published the seminal paper on the solution of mp-MIQP problems using a decomposition based approach, which still represents the state-of-the-art algorithm for this class of problems [6].[2] However, in the paper no comparison procedure was applied, and envelopes of solutions are created.

Since then, several researches have considered mp-MIQP problems: in 2003, Borrelli *et al.* considered the special case of hybrid control and proposed an exhaustive enumeration of the binary variables and solution of the resulting mp-QP problems at each stage [8–10]; a similar strategy was proposed by Alessio and Bemporad [11]. However, it was not until 2011, when Axehill *et al.* considered the reduction of number of solutions per envelope by solving the (potentially non-convex) quadratic programming problem $d_{\min} = \min_{\theta \in CR} \Delta z(\theta)$ and $d_{\max} = \max_{\theta \in CR} \Delta z(\theta)$. If $d_{\min} > 0$ or $d_{\max} < 0$, then one of the objective functions could not be optimal anywhere in the critical region CR considered and thus could be discarded [12, 13].

Following a similar route, in 2014 Oberdieck *et al.* further reduced the number of solutions per envelope by utilizing McCormick relaxations of $\Delta z(\theta)$ [14]. However, it has been shown that this approach increases the overall number of critical regions due to the extra partitions introduced in the parameter space [15]. Lastly, in 2015 Oberdieck and Pistikopoulos developed the first exact solution strategy that circumvents altogether the use of envelopes of solutions and thus yields quadratically constrained critical regions [15].

Additionally, in 2015 Fuchs *et al.* enabled the use of the post-processing strategies of mp-QP problems by lifting the parameter space after the solution has been obtained [16], while in the same year Han and Chen investigated the continuity property of mp-MIQP problems [17].

2 Note that already in 1999 Dua and Pistikopoulos published the global-optimization based approach for the case of general multi-parametric mixed-integer nonlinear programming problems [7].

References

1 Radke, M.A. (1975) *Sensitivity analysis in discrete optimization*, Ph.D. thesis, Caltech, Los Angeles, CA.

2 McBride, R.D. and Yormark, J.S. (1980) Finding all solutions for a class of parametric quadratic integer programming problems. *Management Science*, 26 (8), 784–795, doi: 10.1287/mnsc.26.8.784. URL http://dx.doi.org/10.1287/mnsc.26.8.784.

3 Bank, B. and Hansel, R. (1984) Stability of mixed-integer quadratic programming problems, in *Sensitivity, stability and parametric analysis, Mathematical programming studies*, vol. 21 (ed. A.V. Fiacco), Springer-Verlag, Berlin, Heidelberg, pp. 1–17, doi: 10.1007/BFb0121208. URL http://dx.doi.org/10.1007/BFb0121208.

4 Skorin-Kapov, J. and Granot, F. (1987) Non-linear integer programming: sensitivity analysis for branch and bound. *Operations Research Letters*, 6 (6), 269–274, doi: 10.1016/0167-6377(87)90041-1. URL http://www.sciencedirect.com/science/article/pii/0167637787900411.

5 Greenberg, H.J. (1998) An annotated bibliography for post-solution analysis in mixed integer programming and combinatorial optimization, in (eds. D.L. Woodruff)*Advances in computational and stochastic optimization, logic programming, and heuristic search, Interfaces in computer science and operations research*, Springer US, Boston, MA, pp. 97–147, doi: 10.1007/978-1-4757-2807-1_4. URL http://dx.doi.org/10.1007/978-1-4757-2807-1_4.

6 Dua, V., Bozinis, N.A., and Pistikopoulos, E.N. (2002) A multiparametric programming approach for mixed-integer quadratic engineering problems. *Computers and Chemical Engineering*, 26 (4–5), 715–733, doi: 10.1016/S0098-1354(01)00797-9. URL http://www.sciencedirect.com/science/article/pii/S0098135401007979.

7 Dua, V. and Pistikopoulos, E.N. (1999) Algorithms for the solution of multiparametric mixed-integer nonlinear optimization problems. *Industrial and Engineering Chemistry Research*, 38 (10), 3976–3987, doi: 10.1021/ie980792u.

8 Borrelli, F., Baotic, M., Bemporad, A., and Morari, M. (2003) An efficient algorithm for computing the state feedback optimal control law for discrete time hybrid systems, in *Proceedings of the 2003 American Control Conference, 2003*, vol. 6, pp. 4717–4722, doi: 10.1109/ACC.2003.1242468.

9 Borrelli, F. (2003) *Constrained optimal control of linear and hybrid systems, Lecture notes in control and information sciences*, vol. 290, Springer, Berlin and New York.

10 Borrelli, F., Baotić, M., Bemporad, A., and Morari, M. (2005) Dynamic programming for constrained optimal control of discrete-time linear hybrid systems. *Automatica*, 41 (10), 1709–1721, doi: 10.1016/j.automatica.2005.04.017. URL http://www.sciencedirect.com/science/article/pii/S0005109805001524.

11 Alessio, A. and Bemporad, A. (2006) Feasible mode enumeration and cost comparison for explicit quadratic model predictive control of hybrid systems, in *Proceedings of 2nd IFAC Conference on Analysis and Design of Hybrid Systems, Alghero, Italy (7–9 June 2006)*, pp. 302–308.

12 Axehill, D., Besselmann, T., Raimondo, D.M., and Morari, M. (2011) Suboptimal explicit hybrid MPC via branch and bound, in *IFAC World Congress*, Milano. URL http://control.ee.ethz.ch/index.cgi?page=publications;action=details;id=3759.

13 Axehill, D., Besselmann, T., Raimondo, D.M., and Morari, M. (2014) A parametric branch and bound approach to suboptimal explicit hybrid MPC. *Automatica*, 50 (1), 240–246, doi: 10.1016/j.automatica.2013.10.004. URL http://www.sciencedirect.com/science/article/pii/S0005109813004950.

14 Oberdieck, R., Wittmann-Hohlbein, M., and Pistikopoulos, E.N. (2014) A branch and bound method for the solution of multiparametric mixed integer linear programming problems. *Journal of Global Optimization*, 59 (2–3), 527–543, doi: 10.1007/s10898-014-0143-9. URL http://dx.doi.org/10.1007/s10898-014-0143-9.

15 Oberdieck, R. and Pistikopoulos, E.N. (2015) Explicit hybrid model-predictive control: the exact solution. *Automatica*, 58, 152–159, doi: 10.1016/j.automatica.2015.05.021. URL http://www.sciencedirect.com/science/article/pii/S0005109815002277.

16 Fuchs, A., Axehill, D., and Morari, M. (2015) Lifted evaluation of mp-MIQP solutions. *IEEE Transactions on Automatic Control*, 60 (12), 3328–3331, doi: 10.1109/TAC.2015.2417853.

17 Han, Y. and Chen, Z. (2015) Continuity of parametric mixed-integer quadratic programs and its application to stability analysis of two-stage quadratic stochastic programs with mixed-integer recourse. *Optimization*, 64 (9), 1983–1997, doi: 10.1080/02331934.2014.891033. URL http://dx.doi.org/10.1080/02331934.2014.891033.

7

Solution Strategies for mp-MILP and mp-MIQP Problems

In Chapters 5 and 6, mp-MILP and mp-MIQP problems were discussed in detail. In this chapter, the most common algorithms used for the solution of such problems are described, utilizing an underlying common framework. In particular, consider the general mp-MIQP problem:

$$\begin{aligned} z(\theta) = \quad & \underset{x,y}{\text{Minimize}} \quad (Q\omega + H\theta + c)^T\omega \\ & \text{Subject to} \quad Ax + Ey \leq b + F\theta \\ & \qquad A_{eq}x + E_{eq}y = b_{eq} + F_{eq}\theta \\ & \qquad \theta \in \Theta := \{\theta \in \mathbb{R}^q | \mathrm{CR}_A\theta \leq \mathrm{CR}_b\} \\ & \qquad x \in \mathbb{R}^n, \ y \in \{0,1\}^p, \ \omega = [x^T \ y^T]^T, \end{aligned} \tag{7.1}$$

where $Q \in \mathbb{R}^{(n+p)\times(n+p)}$ is symmetric positive definite, $H \in \mathbb{R}^{(n+p)\times q}$, $c \in \mathbb{R}^{(n+p)}$, $A \in \mathbb{R}^{m\times n}$, $E \in \mathbb{R}^{m\times p}$, $b \in \mathbb{R}^m$, $F \in \mathbb{R}^{m\times q}$, $A_{eq} \in \mathbb{R}^{s\times n}$, $E_{eq} \in \mathbb{R}^{s\times p}$, $b_{eq} \in \mathbb{R}^s$, $F_{eq} \in \mathbb{R}^{s\times q}$, $\mathrm{CR}_A \in \mathbb{R}^{r\times q}$, $\mathrm{CR}_b \in \mathbb{R}^r$, $Ax \leq (b - E\bar{y})$ is a compact polytope for all feasible $\bar{y} \in \{0,1\}^p$. Note that problem (7.1) will be used for the general description of the algorithms for mp-MILP and mp-MIQP problems, and the mp-MILP case is only discussed separately if appropriate.

Multi-parametric Optimization and Control, First Edition.
Efstratios N. Pistikopoulos, Nikolaos A. Diangelakis, and Richard Oberdieck.

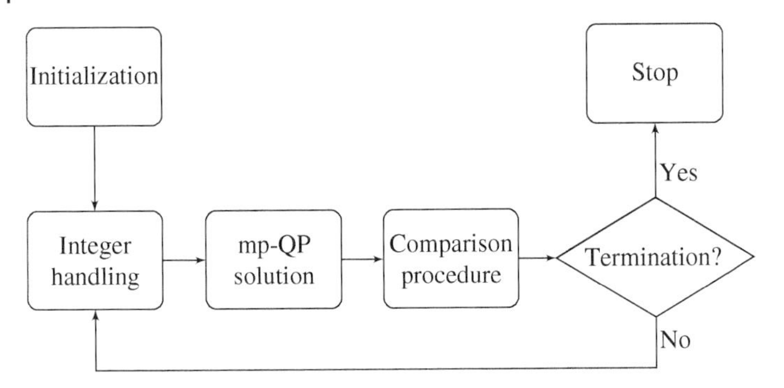

Figure 7.1 The general framework for the solution of mp-MIQP problems.

7.1 General Framework

In general, all strategies for the solution of mp-MIQP problems of type (7.1) that have been presented in the open literature adhere to the following framework (see Figure 7.1):

Initialization: The algorithm is initialized.

Integer handling: A candidate integer solution is found, which is fixed in the original problem, thus transforming it into an mp-QP problem. The three options to find a suitable integer candidate are (i) global optimization, (ii) exhaustive enumeration, and (iii) branch-and-bound. Note that (i) is also referred to as decomposition-based approach.

mp-QP solution: The created mp-QP problem is solved using the strategies presented in Chapter 4.

Comparison procedure: The objective function values of the mp-QP problem and the upper bound in the critical region considered are compared against each other to form the new, tighter upper bound. The four comparison procedures presented are (i) no comparison of the objective function, (ii) comparison of the objective function over the entire critical region considered, (iii) linearization of the nonlinearities in the objective function using McCormick relaxations, and (iv) calculation of the exact solution via piecewise outer approximation of quadratically constrained critical regions. Note that the approaches (i)–(iii) result in envelopes of solutions, i.e. more than one solution is associated with one critical region.

Termination: The algorithm terminates if there are no more regions or potential integer solutions to be explored.

As the initialization procedures and termination criteria exclusively depend on the integer handling strategy, the three different strategies will be discussed in detail in Sections 7.2, 7.3, and 7.4. The different comparison procedures will be explained in detail in Section 7.5.

7.2 Global Optimization

Consider problem (7.1). Then it is possible to find the best possible value of $z(\theta)$, i.e. $\min_\theta(z(\theta))$, by treating θ as an optimization variable. Equally, given an upper bound $\hat{z}(\theta)$, it is possible to determine whether a better solution exists over a given parameter space by inserting $z(\theta) \leq \hat{z}(\theta)$ as a constraint in (7.1), where θ is treated as an optimization variable. These considerations have given rise to the following strategy:

(0) Set $\mathcal{N} \leftarrow \{(\Theta, \infty)\}$ and $S \leftarrow \emptyset$, where $\mathcal{N}$ denotes the list of regions still under consideration, combined with the current best upper bound over that region[1] and S denotes the set of regions for which the optimal solution has been determined.

(1) If $\mathcal{N} = \emptyset$, then the algorithm terminates. Otherwise, a region $\hat{P}$ and its corresponding current best upper bound $\hat{z}(\theta)$ is chosen from $\mathcal{N}$. Following the strategy described earlier, problem (7.1) is modified as follows:

  (C1) The parameters are treated as optimization variables, and restricted to $\hat{P}$, i.e. $\theta \in \hat{P}$.

  (C2) The integer solution of the already existing solution is excluded.

  (C3) It is enforced that any new solution must be better than the already existing one.

1 For clarity purposes, only the objective function value $\hat{z}(\theta)$ is reported in $\mathcal{N}$. However, the values of the optimization variables, $\hat{x}(\theta)$ and $\hat{y}$, are also stored.

This gives rise to the following mixed-integer nonlinear programming (MINLP) problem:

$$
\begin{aligned}
z_{\text{global}} = \underset{x,y,\theta}{\text{minimize}} \;& (Q\omega + H\theta + c)^T\omega \\
\text{subject to} \;& Ax + Ey \leq b + F\theta \\
& A_{\text{eq}}x + E_{\text{eq}}y = b_{\text{eq}} + F_{\text{eq}}\theta \\
& \sum_{j\in J} y_j - \sum_{j\in T} y_j \leq |J| - 1 && \text{(C2)} \\
& (Q\omega + H\theta + c)^T\omega - \hat{z}(\theta) \leq 0 && \text{(C3)} \\
& \theta \in \hat{P}, && \text{(C1)} \\
& x \in \mathbb{R}^n,\; y \in \{0,1\}^p,\; \omega = [x^T\; y^T]^T,
\end{aligned}
\tag{7.2}
$$

where J_i and T_i are the sets containing the indices of the binary variables $\hat{y}$ associated with the upper bound $\hat{z}(\theta)$ that attain the value 0 and 1, respectively, i.e.

$$J = \{j|\hat{y}_j = 1\} \tag{7.3a}$$

$$T = \{j|\hat{y}_j = 0\}. \tag{7.3b}$$

In the case where the current best upper bound is ∞, constraints (C2) and (C3) are omitted.[2] If problem (7.2) is infeasible, then the optimal solution has been found, i.e. $S \leftarrow \{\hat{P}, \hat{z}(\theta)\}$, and the algorithm repeats Step 1. Otherwise, the algorithm proceeds to Step 2.

(2) The binary solution obtained from the MINLP problem is fixed in problem (7.1), resulting in an mp-QP problem, which is solved using the approaches presented in Chapter 4. The resulting solution $\{\Theta_f, z(\theta)\}$ is composed of a set of non-overlapping critical regions CR_i and corresponding optimal objective function values $z_i(\theta)$.[3] Each one of these critical regions with its associated solution is compared with $\{\hat{P}, \hat{z}(\theta)\}$, resulting in a new best upper bound $\{P, z^*(\theta)\}$ (for the different comparison procedures, see Section 7.5). This new upper bound is added to the list of regions and bounds to explore, i.e. $\mathcal{N} \leftarrow \{P, z^*(\theta)\}$, and the algorithm returns to Step 1. It is important to note that $\{P, z^*(\theta)\}$ may be a set of (non-convex) regions.

2 The constraints (C2) and (C3) are also referred to as integer and parametric cut, respectively.

3 This nomenclature is taken from Theorem 3.1. Note that $\Theta_f \subseteq \hat{P}$.

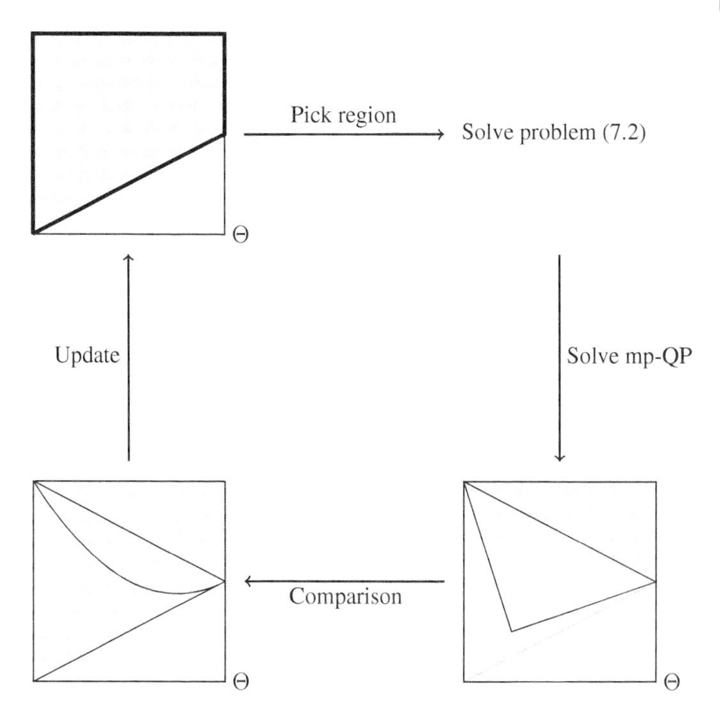

Figure 7.2 A schematic representation of the global optimization approach for the solution of mp-MIQP problems. Starting from a given upper bound, a region is selected and problem (7.2) is solved. Then, the obtained integer solution is substituted into problem (7.1), and the resulting mp-QP problem is solved. This solution is compared with the best upper bound, yielding a new, refined upper bound. This algorithm iterates until no further improvements can be made.

A schematic representation of the global optimization approach is shown in Figure 7.2.

7.2.1 Introducing Suboptimality

Solving mp-MIQP problems can be computationally very demanding. Therefore, there may be cases where calculating a suboptimal solution is sufficient. For the global optimization approach, two aspects have been considered. First, it may be of relevance to know the extent of the entire feasible space of problem (7.1). This can be achieved by performing Step 1 of the iteration only until $\mathcal{N}$ does not contain any elements with $\hat{z}(\theta) = \infty$, i.e. the feasible space of the problem has been calculated. This can be very relevant for feasibility studies or multi-stage optimization problems, where constraint sets are propagated.

Second, suboptimality in a classical sense may also be calculated by introducing the absolute suboptimality gap ϵ in (C3) in problem (7.2), i.e.

$$(Q\omega + H\theta + c)^T\omega - \hat{z}(\theta) + \epsilon \leq 0. \tag{7.4}$$

The larger ϵ becomes, the better the improvement coming from the new candidate solution has to be.

7.3 Branch-and-Bound

Branch-and-bound strategies are very widely used for the solution of mixed-integer programming problems. In essence, the idea is to relax binary variables to bound continuous variables, i.e. $y \in \{0, 1\}^p \rightarrow \overline{y} \in [0, 1]^p$, and the solution of this relaxed problem provides a lower bound to the original problem. From the solution of the relaxed problem, one binary variable is selected and fixed to 0 and to 1. This results in two new relaxed problems, which are then solved and branched further, each time fixing a binary variable either to 0 or 1. The branching only stops if one of the following conditions is met:

(C1) The problem at a given node is infeasible.

(C2) The solution of a problem at a given node is larger than a known upper bound.

(C3) The solution of a problem at a given node is an integer solution.

The power of the branch-and-bound approach derives from the fact that if any of these conditions is met, all the leaves of this node are also discarded.[4] A schematic representation of an example tree is shown in Figure 7.3.

4 In branch-and-bound nomenclature, the tree of all possible realizations consists of many individual nodes which are linked together by branches. All nodes which follow a given node are called its leaves.

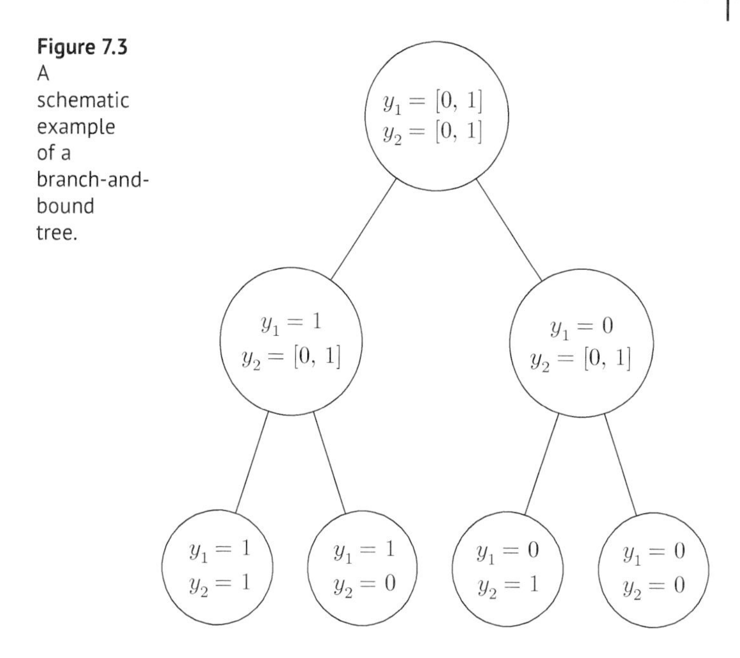

Figure 7.3 A schematic example of a branch-and-bound tree.

Remark 7.1 It is beyond the scope of this book to discuss branch-and-bound algorithms in detail. For a great resource on the topic, the reader is referred to Floudas' textbook on nonlinear and mixed-integer optimization [1].

Due to the wide use of branch-and-bound algorithms in mixed-integer programming, it is natural to translate this algorithm to multi-parametric programming. The key challenge is thereby that (i) the problems to be solved at each node are not LP or QP problems, but mp-LP and mp-QP problems and (ii) the previously mentioned conditions need to be verified for n-dimensional spaces. Thus, the algorithm to solve mp-MIQP problems using a branch-and-bound approach is conceptually identical to the classical branch-and-bound algorithm:

(0) Set $\mathcal{N} \leftarrow \{\Theta, K\}$ and $S \leftarrow \emptyset$, where $\mathcal{N}$ denotes the nodes, represented by the considered feasible parameter space[5] and the

5 This is an element which is not present in the classical branch-and-bound algorithm, where the optimization problem is only solved for a single point.

specific relaxation of the binary variables to be considered and $\mathcal{S}$ denotes the set of regions for which the optimal solution has been determined. Note that the relaxation is defined by (i) the indices of the relaxed variables, (ii) the indices of the fixed variables, and (iii) the values of the fixed variables, i.e. $K = \{k, \emptyset, \emptyset\}$, where $k = 1, \ldots, p$.

(1) If $\mathcal{N} = \emptyset$, then the algorithm terminates. Otherwise, a region $\overline{P}$ and a specific relaxation of binary variables $K = \{k, j, t\}$ is chosen from $\mathcal{N}$. Using this relaxation in problem (7.1), this yields the following mp-QP problem:

$$\begin{aligned} z(\theta) = \quad & \underset{x,\overline{y}}{\text{Minimize}} \quad (Q\omega + H\theta + c)^T\omega \\ & \text{Subject to} \quad Ax + E\overline{y} \le b + F\theta \\ & \qquad A_{\text{eq}}x + E_{\text{eq}}\overline{y} = b_{\text{eq}} + F_{\text{eq}}\theta \\ & \qquad \theta \in \overline{P}, \ x \in \mathbb{R}^n, \\ & \qquad \overline{y}_k \in [0,1]^{|k|}, \ \overline{y}_j = t, \ \omega = [x^T \ \overline{y}^T]^T. \end{aligned} \tag{7.5}$$

If problem (7.5) is infeasible, the algorithm repeats Step 1. If problem (7.5) is feasible, there are several things to be observed about problem (7.5):

- The solution of problem (7.5) is a set of polytopes, the union of which may only be a subset of $\overline{P}$. Furthermore, each polytope features its own solution.
- Depending on the value of H, problem (7.5) might be non-convex. This would preclude the direct use of any strategies presented in Chapter 4.
- The region $\overline{P}$ has to be a polytope in order for the mp-QP algorithms from Chapter 4 to be applicable. Therefore, if non-polytopic regions are considered, outer polytopic approximations of those regions have to be constructed.

(2) Let the solution of problem (7.5) be given by CR_i as the i-th critical region and $x_i(\theta), \overline{y}_i(\theta)$, and $z_i(\theta)$ as the parametric solutions of the continuous optimization variables, relaxed binary variables, and objective function, respectively. Then, based on the previously discussed conditions, each one of these solutions is analyzed separately according to the following criteria:

- Does there exist at least one point θ^* for which the obtained solution is better than the currently best upper bound, i.e. $\exists\theta \in \text{CR}_i$ s.t. $z_i(\theta) < \hat{z}(\theta)$, where $\hat{z}(\theta)$ is the currently best upper bound for problem (7.1) available from S[6] ? This question is answered by comparing $z_i(\theta)$ with $\hat{z}(\theta)$ for $\theta \in \text{CR}_i$. Note though that $\hat{z}(\theta)$ is a set of potentially non-polytopic regions, each of which is associated with a separate solution. These details will be discussed in detail in Chapter 7.5.

- Is $\bar{y}_{i,k}(\theta)$ an integer solution, i.e. is $\bar{y}_{i,k}(\theta) \in \{0, 1\}^{|k|}$?

Let Q be the set of regions where $z_i(\theta) < \hat{z}(\theta)$. Then, if there exists at least a point for which the solution in a given critical region is better, i.e. if $Q \neq \emptyset$, and if the solution is non-integer, then one element of $\bar{y}_k$ is fixed to 0 and 1, i.e. let i be the index of the element to be fixed, then $\mathcal{N} \leftarrow \{\text{CR}_i, K_0\}$ and $\mathcal{N} \leftarrow \{\text{CR}_i, K_1\}$ with $K_0 = \{k\backslash i, j \cup \{i\}, t \cup \{0\}\}$ and $K_1 = \{k\backslash i, j \cup \{i\}, t \cup \{1\}\}$, respectively.

Conversely, $Q \neq \emptyset$, and the solution is integer, then the solution is added to S, i.e. $S \leftarrow \{Q, z_i(\theta)\}$.[7] After all critical regions CR_i have been considered, the algorithm returns to Step 1.

7.4 Exhaustive Enumeration

Especially for smaller problems, it may be computationally feasible to exhaustively enumerate all possible integer combinations, and solve the resulting mp-QP problems. After comparing all these solutions, the optimal solution is obtained. It can therefore be viewed as a special case of the branch-and-bound approach, where instead of starting at the root node, only the node with $y \in \{0, 1\}^p$ is considered. This translates into the following algorithmic structure:

(0) Set $\mathcal{N} = Y$ and $S \leftarrow \emptyset$, where $\mathcal{N}$ denotes the nodes to be considered and S denotes the set of regions for which the optimal solution has been determined. The set Y denotes all possible combinations of $y \in \{0, 1\}^p$, i.e. $Y = \{y | y \in \{0, 1\}^p\}$.

---

6 It is a matter of definition whether $z_i(\theta) < \hat{z}(\theta)$ or $z_i(\theta) \leq \hat{z}(\theta)$ is used as a criterion in this case, i.e. whether less-than or less-or-equal should be used for the decision process. In this book, the less-than notation is used.

7 For clarity purposes, only the objective function value $\hat{z}(\theta)$ is reported in $\mathcal{N}$. However, the values of the optimization variables, $\hat{x}(\theta)$ and $\hat{y}$, are also stored.

(1) If $\mathcal{N} = \emptyset$, then go to Step 2. Otherwise, pop an element of $\hat{y} \in \mathcal{N}$, and solve the following mp-QP problem:

$$\begin{aligned} z(\theta) = \quad & \underset{x}{\text{Minimize}} \quad (Q\omega + H\theta + c)^T\omega \\ & \text{Subject to} \quad Ax + E\hat{y} \leq b + F\theta \\ & \qquad A_{eq}x + E_{eq}\hat{y} = b_{eq} + F_{eq}\theta \\ & \qquad \theta \in \Theta, \ x \in \mathbb{R}^n, \ \omega = [x^T \ \hat{y}^T]^T. \end{aligned} \tag{7.6}$$

If problem (7.6) is feasible, add the solution to $\mathcal{S}$. Repeat Step 1.

(2) Compare all solutions in $\mathcal{S}$ against each other to obtain the optimal solution.[8]

7.5 The Comparison Procedure

Given a critical region CR, an upper bound $\hat{z}(\theta)$, and a solution $z(\theta)$, the aim is to compare these solutions over CR to obtain a new, tighter upper bound $z^*(\theta)$, i.e.

$$z^*(\theta) = \min \ \{\hat{z}(\theta), z(\theta)\}. \tag{7.7}$$

Remark 7.2 Here, we assume that $\hat{z}(\theta)$ and $z(\theta)$ are quadratic functions defined over CR. This situation can be directly reached from the algorithms described earlier by appropriate calculation of the intersection of critical regions and their associated solutions and upper bounds (see, e.g. Figures 5.3 and 6.3). Also, without loss of generality it was assumed that only one objective function is associated with each critical region, and that no envelope of solutions is present (see Definition 6.1).

While the point-wise comparison of two function values is straightforward, the solution of Eq. (7.7) requires the comparison of the corresponding objective functions over CR, i.e.

$$\Delta z(\theta) = \hat{z}(\theta) - z(\theta) = 0. \tag{7.8}$$

Due to the quadratic nature of the objective functions, $\Delta z(\theta)$ might be non-convex. Within the open literature, four strategies for the

8 The details on the comparison procedure will be discussed in detail in Chapter 7.5.

solution of problem (7.8) have been presented:

No objective function comparison: This approach does not consider Eq. (7.8) at all and stores both solutions, $\hat{z}(\theta)$ and $z(\theta)$, in CR, thus creating an envelope of solutions.

Objective function comparison over the entire CR: In this approach, Eq. (7.8) is solved over the entire critical region CR, i.e. the following (possibly non-convex) quadratic programming problems are solved:

$$\delta_{max} = \underset{\theta \in CR}{\text{Maximize}}\ \Delta z(\theta) \tag{7.9a}$$

$$\delta_{min} = \underset{\theta \in CR}{\text{Minimize}}\ \Delta z(\theta). \tag{7.9b}$$

Note that solving Eq. (7.9) potentially is not straightforward since it may be non-convex. The results of solving Eq. (7.9) allow for the following conclusions:

$$\delta_{max} \leq 0 \rightarrow \quad z(\theta) \geq \hat{z}(\theta) \quad \forall \theta \in CR \tag{7.10a}$$

$$\delta_{min} \geq 0 \rightarrow \quad z(\theta) \leq \hat{z}(\theta) \quad \forall \theta \in CR, \tag{7.10b}$$

which enables an appropriate update of the current best upper bound. If $\delta_{min} < 0$ and $\delta_{max} > 0$, then both solutions are kept and an envelope of solutions is created.

7.5.1 Affine Comparison

Instead of solving Eq. (7.9) over the entire critical region, this approach applies the decision criteria in Eq. (7.10) by creating affine under- and overestimators of $\Delta z(\theta)$. This allows for the creation of additional affine constraints where a certain solution is guaranteed to be optimal. Specifically, Eq. (7.8) can be formulated as

$$\Delta z(\theta) = \hat{z}(\theta) - z^*(\theta) = \theta^T P \theta + f^T \theta + w, \tag{7.11}$$

where $P \in \mathbb{R}^{q \times q}$, $f \in \mathbb{R}^q$, and $w \in \mathbb{R}$. If $P = 0_{q \times q}$, then $\Delta z(\theta)$ is an affine function, the decision criteria from Eq. (7.10) can readily be used, i.e.

$$\begin{cases} CR^1 = CR \cap \Delta z(\theta) \leq 0 \\ CR^2 = CR \cap \Delta z(\theta) \geq 0, \end{cases} \tag{7.12}$$

where CR^1 and CR^2 are polytopes and in CR^1 $z^*(\theta)$ is optimal while in CR^2 $\hat{z}(\theta)$ remains optimal. However, if $P \neq 0_{q\times q}$, then a linear under- and overestimator is created, satisfying

$$g(\theta) = a_u^T\theta + b_u \tag{7.13a}$$

$$h(\theta) = a_o^T\theta + b_o \tag{7.13b}$$

with

$$g(\theta) \leq \Delta z(\theta) \leq h(\theta), \quad \forall \theta \in \text{CR} \tag{7.14}$$

where the subscripts u and o indicate the coefficients of the under- and overestimator, respectively. This enables the definition of the following three critical regions:

$$\begin{cases} \text{CR}^1 = \text{CR} \cap g(\theta) \geq 0 \\ \text{CR}^2 = \text{CR} \cap h(\theta) \leq 0 \\ \text{CR}^3 = \text{CR} \cap g(\theta) \leq 0, h(\theta) \geq 0, \end{cases} \tag{7.15}$$

where in CR^1 $z^*(\theta)$ is optimal while in CR^2 $\hat{z}(\theta)$ remains optimal, and in CR^3 both solutions are stored in an envelope of solutions. Since $h(\theta)$ and $g(\theta)$ are affine functions, CR^1, CR^2, and CR^3 are polytopes.

In order to find $g(\theta)$ and $h(\theta)$, the bilinear terms in the definition of $\Delta z(\theta)$ need to be addressed. For every bilinear term $\theta_i\theta_j$ the McCormick under- and overestimator[9] are created according to the following:

$$\theta_i\theta_j \geq \max\{\theta_j^{\max}\theta_i + \theta_i^{\max}\theta_j - \theta_i^{\max}\theta_j^{\max}, \theta_j^{\min}\theta_i + \theta_i^{\min}\theta_j - \theta_i^{\min}\theta_j^{\min}\} \tag{7.16a}$$

$$\theta_i\theta_j \leq \min\{\theta_j^{\max}\theta_i + \theta_i^{\min}\theta_j - \theta_i^{\min}\theta_j^{\max}, \theta_j^{\min}\theta_i + \theta_i^{\max}\theta_j - \theta_i^{\max}\theta_j^{\min}\} \tag{7.16b}$$

Note that $\theta_i^{\min}$ and $\theta_i^{\max}$ are obtained *via* the solution of linear programming problems.

9 This approach is not limited to the use of McCormick estimators, as any valid linear under- and overestimator can be used.

7.5.2 Exact Comparison

All approaches presented so far may create an envelope of solutions. Thus, they are implicit representations of the solution and for a given parameter realization, an additional point-wise comparison is necessary to identify the optimal solution. Thus, the solutions are not "exact," as exact description of the critical region is not explicitly available in its potentially non-convex form given in Eq. (6.4). Such a representation would enable the assignment of one solution to each region, and consequently an assessment of the impact and meaning of each region.

In this exact approach, the use of envelopes of solutions is completely avoided. Following the solution of the non-convex equation (7.9), the following non-convex critical regions are generated:

$$\begin{cases} \mathrm{CR}^1 = \mathrm{CR} \cap \Delta z(\theta) \leq 0 \\ \mathrm{CR}^2 = \mathrm{CR} \cap \Delta z(\theta) \geq 0, \end{cases} \tag{7.17}$$

where in CR^1 $z^*(\theta)$ is optimal, while in CR^2 $\hat{z}(\theta)$ remains optimal. The issue at this point is the solution of the mp-QP problem at the next iteration, as well as the solution of Eq. (7.9) for a potentially non-convex critical region. These issues are addressed by creating an outer approximation of CR^1 and CR^2, i.e. $\Xi^1 \supseteq \mathrm{CR}^1$ and $\Xi^2 \supseteq \mathrm{CR}^2$, using McCormick relaxations. The mp-QP problems are solved over these outer approximations, thus guaranteeing that the correct solution is obtained. At the comparison procedure, this outer approximation is removed and the original, non-convex region is recovered. However, this may lead to regions, which are non-empty in the outer approximation Ξ but empty in the original non-convex region CR. To illustrate this concept, consider the case where CR in Eq. (7.17) is non-convex. Then, if the mp-QP was solved using the outer approximation $\Xi \supseteq \mathrm{CR}$, Eq. (7.17) is converted to

$$\begin{cases} \hat{\mathrm{CR}}^1 = \Xi \cap \Delta z(\theta) \leq 0 \\ \hat{\mathrm{CR}}^2 = \Xi \cap \Delta z(\theta) \geq 0. \end{cases} \tag{7.18}$$

In order to remove Ξ and introduce the original, non-convex region, we have to identify whether the following equation holds:

$$\{\theta | \theta \in \hat{CR}^1 \cap CR\} \neq \emptyset \tag{7.19}$$

If equation (7.19) holds, then we define:

$$CR^1 = CR \cap \Delta z(\theta) \leq 0 \tag{7.20}$$

and update the solution accordingly. Note that this is done equivalently for $\hat{CR}^2$.

Remark 7.3 Note that the method described here is not the most efficient way to implement this approach. In Chapter 8, we will discuss a more efficient implementation in the MATLAB® environment.

A schematic representation of the different comparison procedures is shown in Figure 7.4.

7.6 Discussion

None of the algorithms presented in this chapter have been proven to improve on the worst-case complexity of solving problem (7.1). However, an analysis of the algorithms reveals a variety of details that tend to make some approaches more suitable than others.

7.6.1 Integer Handling

Global optimization: This is the most common integer handling strategy, as it scales reasonably well with the number of integer variables. In addition, it has been shown that the solution of problem (7.2) does not tend to be the bottleneck in terms of computational performance. Lastly, the application of the comparison procedure is very natural from an algorithmic perspective: the obtained solution is simply compared over the

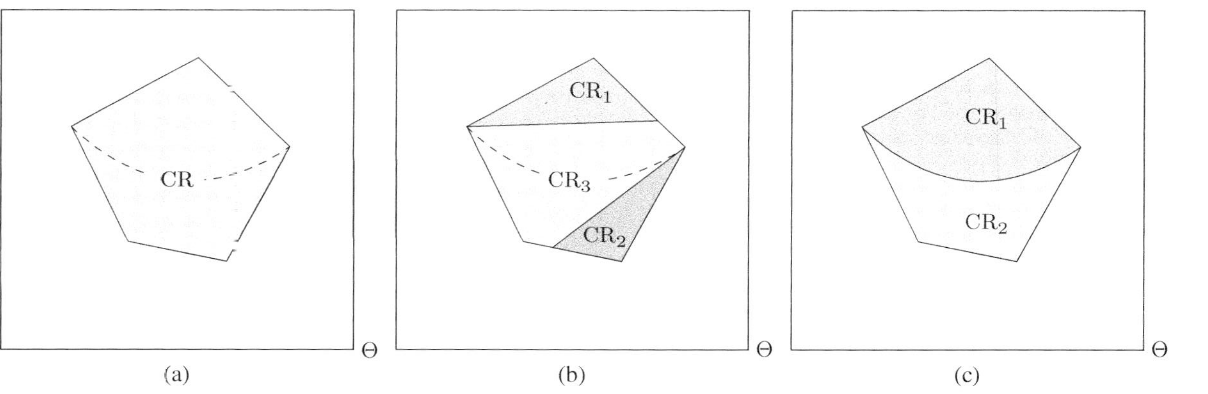

Figure 7.4 A schematic perspective on the different comparison procedures, where the dashed line denotes $\Delta z(\theta) = 0$: in (a), a comparison over the entire critical region is shown. Since both solutions are optimal somewhere in the critical region, no partitioning is taking place, and an envelope of solutions is created. Conversely, in (b), $\Delta z(\theta)$ is linearized, creating three new critical regions. In CR_1 and CR_2, it is guaranteed that only one of the solutions will be optimal, while in CR_3 and envelope of solutions is created. Lastly, in (c) an exact partitioning of the critical region is shown. It explicitly represents the transition between the two solutions, and therefore is rich in structural information. However, it also introduces non-convexity, which increases the complexity of the algorithm and the solution. (a) Comparison over CR. (b) Affine comparison. (c) Exact comparison.

critical region considered as the feasible parameter space in problem (7.2). Therefore, it is not necessary to check which upper bound is valid in the critical region considered (see, for example, Figures 5.3 and 6.3). This is one of the main drawbacks of the branch-and-bound approach (see the following text).

An additional point to make is that the use of the global optimization approach also enables the use of sophisticated global optimization software. For more details on the implementation of the global optimization strategy, please see Chapter 8.

Branch-and-bound: Although the branch-and-bound approach was historically speaking the first one to be described in the literature, it has not been as widely used as the global optimization and exhaustive enumeration approaches. This can be mainly attributed to two factors: first, with an increasing number of integer variables, the number of mp-QP problems to be solved in the approach dramatically increases. This is due to the fact that the fathoming efficiency (i.e. the ability to discard nodes and their leaves) tends to be very low, as it has to be guaranteed that the entire parameter space is covered by one of the three criteria. Thus, many different nodes have to be propagated, reducing the computational efficiency of the algorithm. Second, for each solution found at a given node, it has to be compared with the best upper bound over the entire parameter space. Thus, situations such as shown in Figures 5.3 and 6.3 have to be checked for, which with an increasing number of critical regions, becomes computationally very challenging.

Thus, although the branch-and-bound approach does not require the use of global optimization software, it is at the moment not a widely used approach to solving problems of type (7.1).

Exhaustive enumeration: Although the simplest strategy among the areas considered here, it has still be used in a variety of settings to solve specific problems. In particular when the number of integer variables is small, either by design or imposed by the structure of the problem, this approach has been used as it avoids the solution of problem (7.2). However, its clear limitation is the scalability, as the number of possible combinations increases exponentially.

7.6.2 Comparison Procedure

Before going into the specific discussion, the following remarks are made:

- Comparison procedures have to be assessed regarding (i) the number of solutions per critical region, (ii) the number of partitions created, and (iii) computational effort.
- Without any post-processing (e.g. merging of critical regions with the same solution), the number of critical regions in the final solution is distributed as follows:

$$\begin{aligned}\text{No comparison} &= \text{Comparison over CR}\\ &\leq \text{Exact comparison}\\ &\leq \text{Affine comparison.}\end{aligned}$$

  This is due to the fact that the "no comparison" and "comparison over CR" approaches do not introduce any new partitions, as they simply use the critical regions given by the mp-QP problem. The "exact comparison" performs the exact partitioning to show the solution as is, while the "affine comparison" introduces extra partitions due to the linearization procedure.
- In the case of mp-MILP problems, the "exact comparison" and the "affine comparison" are identical, since P in Eq. (7.11) is zero.

No comparison: This is the easiest strategy when dealing with the comparison, i.e. not dealing with it at all. Due to its simplicity, and consequent ease of implementation, it has been used extensively in the literature. However, its drawback is that with an increasing number of integer variables, the number of solutions per critical region tends to increase as well. Therefore, this procedure is mostly suited for smaller problems, as well as for prototyping. However, as mentioned earlier, the number of critical regions is minimal when no post-processing is applied.

Objective function comparison over the entire CR: This strategy strikes a good balance between the simplicity of the "no comparison" approach and reduction in number of solutions per critical region. In fact, it can be viewed as a more sophisticated version of the "no comparison" procedure, as it does not introduce additional partitions, but checks whether the solutions in the envelope of solutions are in fact optimal at any point in the parameter space. The only concern may be the solution of the non-convex equation (7.9). However, using appropriate global optimization solvers, this problem can be solved sufficiently accurate to enable its use.

Affine comparison: This strategy reduces the number of solutions per critical region even further, however at the price of a significant increase in the number of critical regions. Whether this is warranted depends on the specific problem at hand. Although this increase in number of regions may point towards an increased computational burden, initial computational studies have not supported this statement (see Chapter 8). It may in fact be argued that the increased number of critical regions leads to smaller regions for the mp-QP to be solved in (e.g. in the global optimization approach), which fosters quicker termination.

Exact comparison: Out of all the approaches presented here, this approach is the only one that does not create any envelopes of solutions. However, it has also been shown to have the poorest computational performance (see Chapter 8), a fact that makes its application for the solution of large-scale mp-MIQP problems problematic. However, this approach is uniquely suited for the analysis of specific cases and understanding of the underlying structure of the solution. Therefore, it can be viewed as complementary to the three other approaches presented here.

7.7 Literature Review

As mentioned in Section 4.6,many researchers who have worked on multi-parametric programming have done so by proposing a novel solution technique. In order to be consistent with the chapter, this review will only cover approaches handling multiple parameters.

After the first mp-MILP algorithm was presented in 1997 featuring a branch-and-bound approach combined with the affine comparison, the next 6 years produced much of the literature on the solution of mp-MILP and mp-MIQP problems. After inactivity in

Table 7.1 An overview over the literature on how to solve mp-MILP and mp-MIQP problems, ordered by integer handling strategy and comparison procedure.

	Global optimization	**Branch and bound**	**Exhaustive enumeration**
No comparison	Dua *et al.* [2, 3], Wittmann-Hohlbein and Pistikopoulos [4–7]		Bemporad and coworkers [8], Borrelli *et al.* [9–11]
Comparison over CR		Axehill *et al.* [12, 13]	
Affine comparison		Oberdieck and Pistikopoulos [14]	
Exact comparison	Oberdieck and coworkers [15]	Oberdieck and coworkers [15]	

the field for the next 10 years, there has been some development since 2014 with numerous papers discussing a variety of integer handling and comparison procedures. An overview over the different papers presented is shown in Table 7.1.

The most notable aspect of the development in the literature is that except a small section in [16], no paper has discussed in depth the underlying general framework for the solution of mp-MILP and mp-MIQP problems. In addition, several combinations integer handling and comparison procedure have also not been published. This lack of coverage in the literature is mainly associated with the computational burden commonly associated with general mp-MILP and mp-MIQP problems. As a result, many researchers have focused on the development of solution techniques for their specific problems, such as hybrid explicit model predictive control (see Chapter 10), whose specific structure enables a faster solution of the specific problem at hand. Lastly, the idea of suboptimality, shown in Section 7.2.1, has been discussed by Axehill *et al.* [13] and Habibi *et al.* [17].

However, the recent developments in the POP toolbox have shown, which it is possible to design a computationally efficient and scalable mp-MILP and mp-MIQP solver [18], capable of solving larger mp-MILP and mp-MIQP problems. Together with the ability to design parallel solution algorithms for multi-parametric programming problems, as shown in Oberdieck and Pistikopoulos [19], this significantly enhances the applicability of mp-MILP and mp-MIQP problems.

References

1 Floudas, C.A. (1995) *Nonlinear and mixed-integer optimization: fundamentals and applications, Topics in chemical engineering*, Oxford University Press, New York.

2 Dua, V. and Pistikopoulos, E.N. (2000) An algorithm for the solution of multiparametric mixed integer linear programming problems. *Annals of Operations Research*, 99 (1–4), 123–139, doi: 10.1023/A:1019241000636. URL http://dx.doi.org/10.1023/A %3A1019241000636.

3 Dua, V., Bozinis, N.A., and Pistikopoulos, E.N. (2002) A multi-parametric programming approach for mixed-integer quadratic engineering problems. *Computers and Chemical Engineering*, 26 (4–5), 715–733, doi: 10.1016/S0098-1354(01)00797-9. URL http://www.sciencedirect.com/science/article/pii/S0098135401007979.

4 Wittmann-Hohlbein, M. and Pistikopoulos, E.N. (2012) A two-stage method for the approximate solution of general multiparametric mixed-integer linear programming problems. *Industrial and Engineering Chemistry Research*, 51 (23), 8095–8107, doi: 10.1021/ie201408p.

5 Wittmann-Hohlbein, M. and Pistikopoulos, E.N. (2013) On the global solution of multi-parametric mixed integer linear programming problems. *Journal of Global Optimization*, 57 (1), 51–73, doi: 10.1007/s10898-012-9895-2. URL http://dx.doi.org/10.1007/s10898-012-9895-2.

6 Wittmann-Hohlbein, M. and Pistikopoulos, E.N. (2013) Proactive scheduling of batch processes by a combined robust optimization and multiparametric programming approach. *AIChE Journal*, 59 (11), 4184–4211, doi: 10.1002/aic.14140. URL http://dx.doi.org/10.1002/aic.14140.

7 Wittmann-Hohlbein, M. and Pistikopoulos, E.N. (2014) Approximate solution of mp-MILP problems using piecewise affine relaxation of bilinear terms. *Computers and Chemical Engineering*, 61, 136–155, doi: 10.1016/j.compchemeng.2013.10.009. URL http://www.sciencedirect.com/science/article/pii/S0098135413003311.

8 Bemporad, A., Borrelli, F., and Morari, M. (2000) Piecewise linear optimal controllers for hybrid systems, in *Proceedings of the American Control Conference*, vol. 2, pp. 1190–1194, doi: 10.1109/ACC.2000.876688.

9 Borrelli, F., Baotic, M., Bemporad, A., and Morari, M. (2003) An efficient algorithm for computing the state feedback optimal control law for discrete time hybrid systems, in *Proceedings of the 2003 American Control Conference, 2003*, vol. 6, pp. 4717–4722, doi: 10.1109/ACC.2003.1242468.

10 Borrelli, F. (2003) *Constrained optimal control of linear and hybrid systems, Lecture notes in control and information sciences*, vol. 290, Springer, Berlin and New York.

11 Borrelli, F., Baotić, M., Bemporad, A., and Morari, M. (2005) Dynamic programming for constrained optimal control of discrete-time linear hybrid systems. *Automatica*, 41 (10), 1709–1721, doi: 10.1016/j.automatica.2005.04.017. URL http://www.sciencedirect.com/science/article/pii/S0005109805001524.

12 Axehill, D., Besselmann, T., Raimondo, D.M., and Morari, M. (2011) Suboptimal explicit hybrid MPC via branch and bound, in *IFAC World Congress*, Milano. URL http://control.ee.ethz.ch/index.cgi?page=publications;action=details;id=3759.

13 Axehill, D., Besselmann, T., Raimondo, D.M., and Morari, M. (2014) A parametric branch and bound approach to suboptimal explicit hybrid MPC. *Automatica*, 50 (1), 240–246, doi: 10.1016/j.automatica.2013.10.004. URL http://www.sciencedirect.com/science/article/pii/S0005109813004950.

14 Oberdieck, R., Wittmann-Hohlbein, M., and Pistikopoulos, E.N. (2014) A branch and bound method for the solution of multiparametric mixed integer linear programming problems. *Journal of Global Optimization*, 59 (2–3), 527–543, doi: 10.1007/s10898-014-0143-9. URL http://dx.doi.org/10.1007/s10898-014-0143-9.

15 Oberdieck, R. and Pistikopoulos, E.N. (2015) Explicit hybrid model-predictive control: the exact solution. *Automatica*, 58, 152–159, doi: 10.1016/j.automatica.2015.05.021. URL http://www.sciencedirect.com/science/article/pii/S0005109815002277.

16 Pistikopoulos, E.N., Diangelakis, N.A., Oberdieck, R., Papathanasiou, M.M., Nascu, I., and Sun, M. (2015) PAROC - an integrated framework and software platform for the optimization and advanced model-based control of process systems. *Chemical Engineering Science*, 136, 115–138, doi: 10.1016/j.ces.2015.02.030. URL http://www.sciencedirect.com/science/article/pii/S0009250915001451.

17 Habibi, J., Moshiri, B., Sedigh, A.K., and Morari, M. (2016) Low-complexity control of hybrid systems using approximate multi-parametric MILP. *Automatica*, 63, 292–301, doi: 10.1016/j.automatica.2015.10.032. URL http://www.sciencedirect.com/science/article/pii/S000510981500432X.

18 Oberdieck, R., Diangelakis, N.A., Papathanasiou, M.M., Nascu, I., and Pistikopoulos, E.N. (2016) POP – parametric optimization toolbox. *Industrial and Engineering Chemistry Research*, 55 (33), 8979–8991, doi: 10.1021/acs.iecr.6b01913. URL http://dx.doi.org/10.1021/acs.iecr.6b01913.

19 Oberdieck, R. and Pistikopoulos, E.N. (2016) Parallel computing in multi-parametric programming, in *26th European Symposium on Computer Aided Process Engineering, Computer Aided Chemical Engineering*, vol. 38 (eds Z. Kravanja and M. Bogataj), Elsevier, Portorož, Slovenia, pp. 169–174, doi: 10.1016/B978-0-444-63428-3.50033-3. URL http://www.sciencedirect.com/science/article/pii/B9780444634283500333.

8

Solving Multi-parametric Programming Problems Using MATLAB®

Following the discussion of the different solution strategies in Chapters 4 and 7, this chapter focuses on the implementation of these strategies in MATLAB®. Specifically, it gives an introduction to the Parametric Optimization (POP) toolbox, which has been developed by the authors of this book. It features a set of tailor-made algorithms for the solution of mp-LP, mp-QP, mp-MILP, and mp-MIQP problems as well as a versatile problem generator and a comprehensive problem library. In addition, the toolbox is equipped with an intuitive graphical user interface (GUI), which is seamlessly connected to all functionalities of the toolbox.

It is worth noting that the POP toolbox is not the only software tool available for the solution of multi-parametric programming problems. In particular, the Multi-Parametric Toolbox (MPT), developed jointly by ETH Zurich and the University of Bratislava, is widely used. It was first introduced in 2004 [1], and recently MPT 3.1 has been released [2–5]. It is a complete software tool able to solve multi-parametric programming problems and perform key operations of linear algebraic geometry as well as to design explicit controllers very intuitively. In Section 8.6, the computational performance of POP versus MPT is shown, whenever appropriate.[1]

1 The MPT toolbox currently does not feature mp-MILP and mp-MIQP solvers. Conversely, MPT solves multi-parametric linear complementarity problems (mp-LCP, see Section 9.3), a feature which POP does not support.

Multi-parametric Optimization and Control, First Edition.
Efstratios N. Pistikopoulos, Nikolaos A. Diangelakis, and Richard Oberdieck.

8.1 An Overview over the Functionalities of POP

The POP toolbox consists of three different aspects: problem solution, problem generation, and problem library.

8.2 Problem Solution

8.2.1 Solution of mp-QP Problems

In POP, a step-size based geometrical, a variation of the combinatorial and the connected-graph algorithm has been implemented (see Chapter 4 for details). These are accessible as functions in the Command Window:

```
Solution = Geometrical(problem)
Solution = Combinatorial(problem)
Solution = ConnectedGraph(problem),
```

where `problem` is the structured array containing the mp-LP/mp-QP problem to be solved. Additionally, POP provides an interface with the solver used in MPT:

```
Solution = POPviaMPT(problem).
```

Note that this requires the separate download of the MPT toolbox. Thus, POP features every major solution strategy for mp-LP and mp-QP problems. These have been combined in a single wrapper:

```
Solution = mpQP(problem)
```

8.2.2 Solution of mp-MIQP Problems

In POP, a global optimization algorithm and an exhaustive enumeration algorithm have been implemented featuring the four different comparison procedures discussed in Chapter 7. The solver is available in the Command Window as

```
Solution = mpMIQP(problem)
```

8.2.3 Requirements and Validation

It is possible to use all functionalities of POP using only the built-in functionalities of MATLAB and its toolboxes. However, for speed and stability reasons, the use of commercial tools is encouraged. In particular, POP features links to CPLEX and NAG as LP and QP solvers, as well as CPLEX for the MILP and MIQP problems.

In order to validate the obtained solutions, POP features the function `VerifySolution`, which randomly seeds n points[2] in the parameter space Θ and solves the corresponding deterministic problem. While this does not provide a full certificate of guarantee, it is a strong indicator that a correct solution has been obtained.

8.2.4 Handling of Equality Constraints

For the case of mp-LP and mp-QP problems, equality constraints are simply considered as active constraints of the solution. Conversely, for mp-MILP and mp-MIQP problem, the global optimization problem is solved straight up, as it is expected that the chosen solver is capable of handling such issues. Once a candidate combination of binary variables has been found and fixed, the resulting equality constraints are considered in the mp-LP and mp-QP problem.

8.2.5 Solving Problem (7.2)

In the case the global optimization strategy is chosen for the handling of the integer variables, it is necessary to solve the MINLP problem (7.2), which features potentially non-convex quadratic

2 The default value is 5000.

constraints. As CPLEX only provides MILP and MIQP solvers, the quadratic constraints in problem (7.2) are underestimated using a suitable set of McCormick estimators (see Eq. (7.16)). Note that this guarantees correct execution of the algorithm. However, if no comparison procedure is employed, then the number of solutions per critical region might be higher than in the case where a MINLP solver is used.

8.3 Problem Generation

The aim is to generate random, feasible problems with suitably defined constraints such that different active sets become optimal in different parts of the parameter space, thus resulting in a partitioning of the parameter space into several critical regions. Consider the general mp-QP problem:

$$
\begin{aligned}
z(\theta) = \underset{x}{\text{Minimize}}\ & \left(\tfrac{1}{2}Qx + H\theta + c\right)^T x \\
\text{Subject to}\ & Ax \leq b + F\theta \\
& A_{eq}x = b_{eq} + F_{eq}\theta \\
& \theta \in \Theta := \{\theta \in \mathbb{R}^q | CR_A\theta \leq CR_b\} \\
& x \in \mathbb{R}^n,
\end{aligned}
\tag{8.1}
$$

where $Q \in \mathbb{R}^{n\times n}$ is symmetric positive definite, $c \in \mathbb{R}^n$, $A \in \mathbb{R}^{m\times n}$, $b \in \mathbb{R}^m$, $A_{eq} \in \mathbb{R}^{s\times n}$, $b_{eq} \in \mathbb{R}^s$, $Ax \leq b$ is a compact polytope, A_{eq} is assumed to have full rank,[3] $H \in \mathbb{R}^{n\times q}$, $F \in \mathbb{R}^{m\times q}$, $F_{eq} \in \mathbb{R}^{s\times q}$, $CR_A \in \mathbb{R}^{r\times q}$, $CR_b \in \mathbb{R}^r$, and Θ is a compact polytope. The problem generation of a mp-QP of type (8.1) the development of such a generator can be decomposed into the following steps:

Step 1 – *Objective Function*: In order to define the objective function, Q, H, and c according to problem (8.1) need to be defined. While for H and c no specific criterion apply, Q needs to be symmetric positive definite. This is achieved by randomly generating a diagonal matrix featuring positive entries.

---

3 If A_{eq} does not have full rank, it is always possible to find an equivalent matrix A'_{eq} with a reduced number of rows which has full rank.

Step 2 – Constraints: The two criteria for the generation of constraints for multi-parametric programming problems are (i) feasibility and (ii) tightness in the sense that different solutions should be optimal in different parts of the parameter space. Furthermore, any set of linear constraints can be written as a set of matrices. Thus, generating random constraints is equivalent to generating random rows for the constraint matrices. As a result, the problem generator considers each row of the constraints of problem (8.1) individually and assesses (i) how many non-zero entries there should be for A, b, and F, respectively, and (ii) what scale the constraints should be on. This enables the assignment of a random value to each entry of A, b, and F. Note that this principle also applies to mp-LP, mp-MILP, and mp-MIQP problems, respectively. In addition, note that all the parameters parameters, which define, e.g. how many constraints are generated and to what scale they adhere to, are randomly generated in order to minimize the impact of structural bias in the generation process.

Within POP, the problem generator is accessible from the Command Window as

```
problem = ProblemGenerator(Type,Size,options)
```

where `Type` is `''mpLP''`, `''mpQP''`, `''mpMILP''`, or `''mpMIQP,''` and `Size` is a structured array featuring the desired dimensions of the optimization variables, parameters, and constraints. Additionally, the `options` input specifies settings that are discussed in detail in the User Manual. In particular, it is possible to generate more than one problem directly, which enables the seamless generation of problem libraries and test sets.

8.4 Problem Library

The third key feature of POP is its problem library, currently featuring the four randomly generated test sets "POP_mpLP1," "POP_mpQP1," "POP_mpMILP1," and "POP_mpMIQP1" containing 100 randomly generated mp-LP, mp-QP, mp-MILP, and

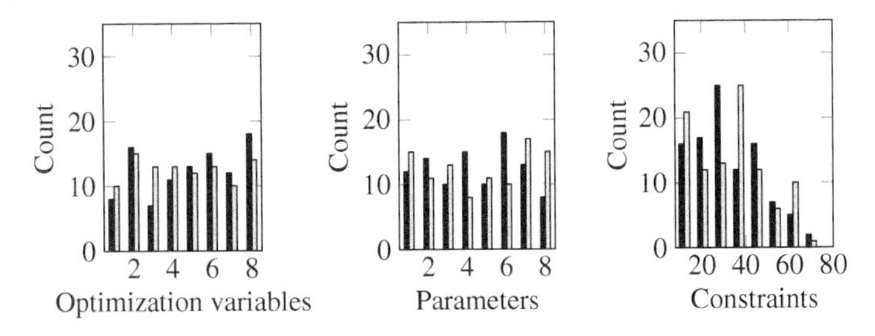

Figure 8.1 The problem statistics of the test sets "POP_mpLP1" and "POP_mpQP1."

mp-MIQP problems respectively (see Figures 8.1 and 8.2). These problem libraries are used to analyze the performance of the different solvers and options available in POP. These test problems represent to our knowledge the first ever comprehensive library of test problems in multi-parametric programming.

Within POP, each problem is stored in the folder "Library," which contains a folder for each test set, which in return contains all the individual problems as ".mat" files. These files can be loaded into MATLAB and the corresponding problem can be solved. Additionally, it is possible to use the GUI (see next section), to perform statistical analysis as well as to create customized test sets which can be exported and solved directly.

8.4.1 Merits and Shortcomings of The Problem Library

The aim of the POP toolbox is not only to provide the means to solve multi-parametric programming problems, but also in general to advance the computational side of multi-parametric programming solvers beyond their description and solution of some test cases. For this purpose, it is vital to create test beds where new and old algorithms can be compared against each other. The larger this basis of test cases is, the more efficient and robust will the implementations of these algorithms be. The problem library included in POP is the first step towards this direction, as it provides 100 problems for each of the four major problem classes. It therefore enables the access to larger quantities of data for solver performance and as a result the inference of bottlenecks in algorithms and comparisons of different solvers.

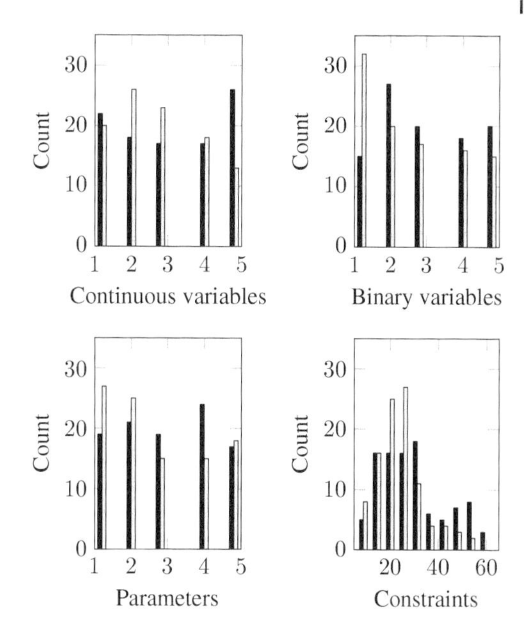

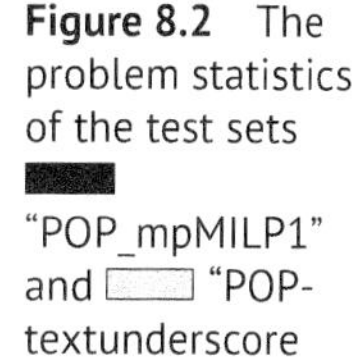

Figure 8.2 The problem statistics of the test sets "POP_mpMILP1" and "POP-textunderscore mpMIQP1."

However, these problems are randomly generated using the problem generator of the POP toolbox. This means that the problems themselves are not based on real-world applications. Thus, the problem library in its current form does not give any information as to what algorithm is more appropriate for an MPC or scheduling application, and therefore conclusions drawn from the results of the problem library should be taken as suggestive and not definitive. In future, the aim is to vastly expand the problem library and automate the benchmarking to an extent that enables the readily available testing of any new implementation.

8.5 Graphical User Interface (GUI)

In order to facilitate its use, POP is equipped with a GUI, which can be launched from the Command Window using

```
POP
```

It enables direct access to the different functions of POP including post-processing and exporting automatically generated code. The main screens of the interface are shown in Figure 8.3, i.e. the welcome screen, and the solver, library, and generator interfaces. Note that in order to maintain a user-friendly approach, some of the options available in POP are set to defaults when the interface is used. More information on the interface can be found in the User Manual.

8.6 Computational Performance for Test Sets

Remark 8.1 The time out of the algorithms are set to 600 seconds, i.e. 10 minutes. Thus, the percentage of problems solved is directly linked to the ability of the algorithm to solve the problem given this time constraint. Note that this setting is user-defined and completely arbitrary.

8.6.1 Continuous Problems

In Figure 8.4, the time versus the percentage of problems of the test set solved is shown for "POP_mpLP1" and "POP_mpQP1." Additionally, Figures 8.5 and 8.6 highlight the computational effort spent on different aspects of each algorithm for the test sets "POP_mpLP1" and "POP_mpQP1," respectively. For the geometrical algorithm, the three aspects considered are (i) solution of the QP problem, (ii) removal of redundant constraints, and (iii) identification of a new point θ_0. For the combinatorial and the connected-graph algorithm, the different aspects are (i) validation whether the selected active set was already considered or can be discarded as infeasible, (ii) establishing feasibility, and (iii) establishing optimality.

8.6.2 Mixed-integer Problems

In Figure 8.7, the time versus the percentage of problems of the test set solved for "POP_mpMILP1" and "POP_mpMIQP1" is shown.

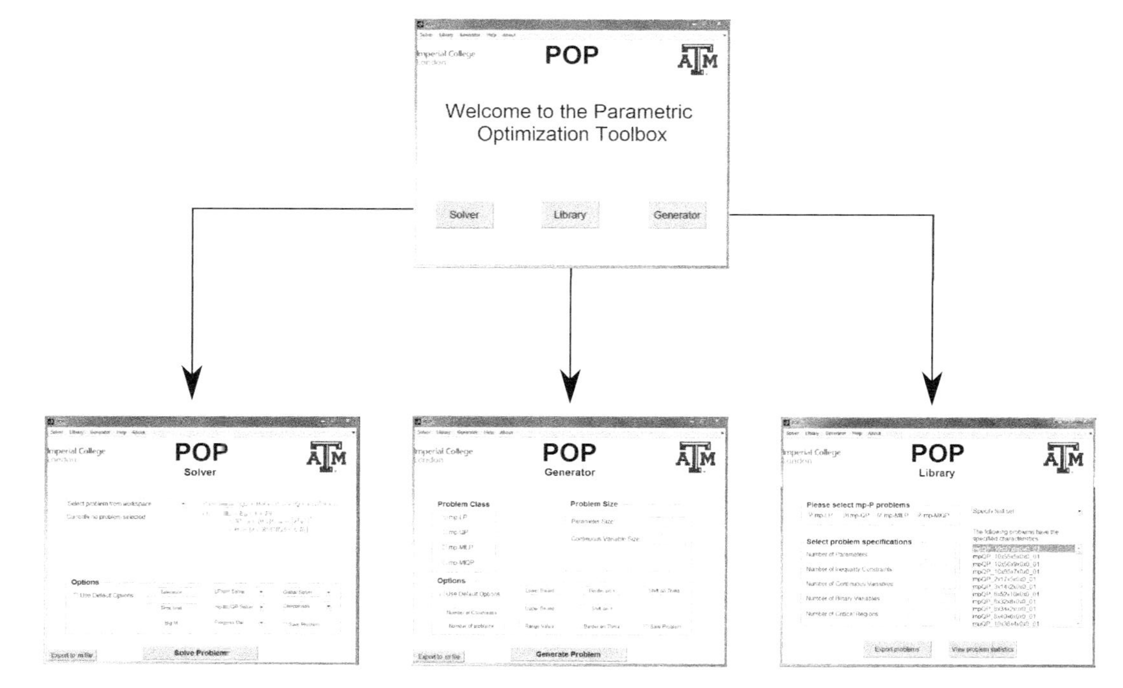

Figure 8.3 The structure of the graphical user interface (GUI) of POP.

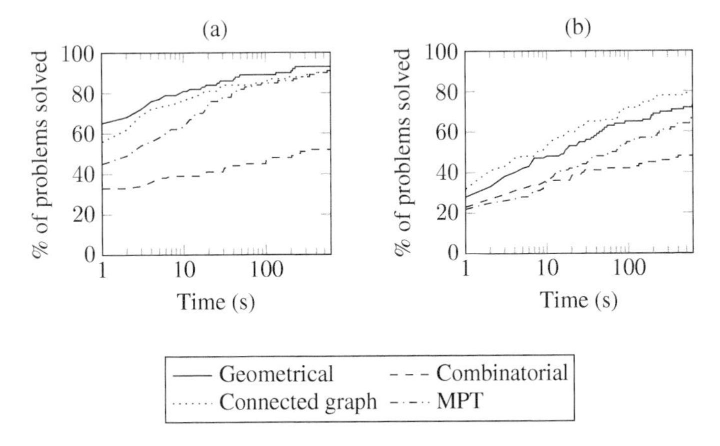

Figure 8.4 The performance of the geometrical, combinatorial, and connected-graph algorithm as well as MPT v3.1 over the test sets (a) "POP_mpLP1" and (b) "POP_mpQP1."

Additionally, Figures 8.8 and 8.9 highlight the computational effort spent on the different aspects of the algorithm, namely, the integer handling, the mp-QP solution, and the comparison procedure for the different comparison procedures considered, i.e.

None: No comparison procedure is used.
MinMax: The sole solution of Eq. (7.9)
Affine: The linearization of $\Delta z(\theta)$
Exact: The calculation of the exact solution for mp-MIQP problems. Note that for mp-MILP problems, the "Affine" comparison already yields the exact solution (i.e. no envelope of solutions) due to the linear nature of the objective function.

8.7 Discussion

The solution of the test set problems highlights the full capabilities of POP. Based on the random problem generator, comprehensive test sets are developed, which are used to compare different algorithms and investigate their computational behavior. This enables the identification of areas for future developments as well as an

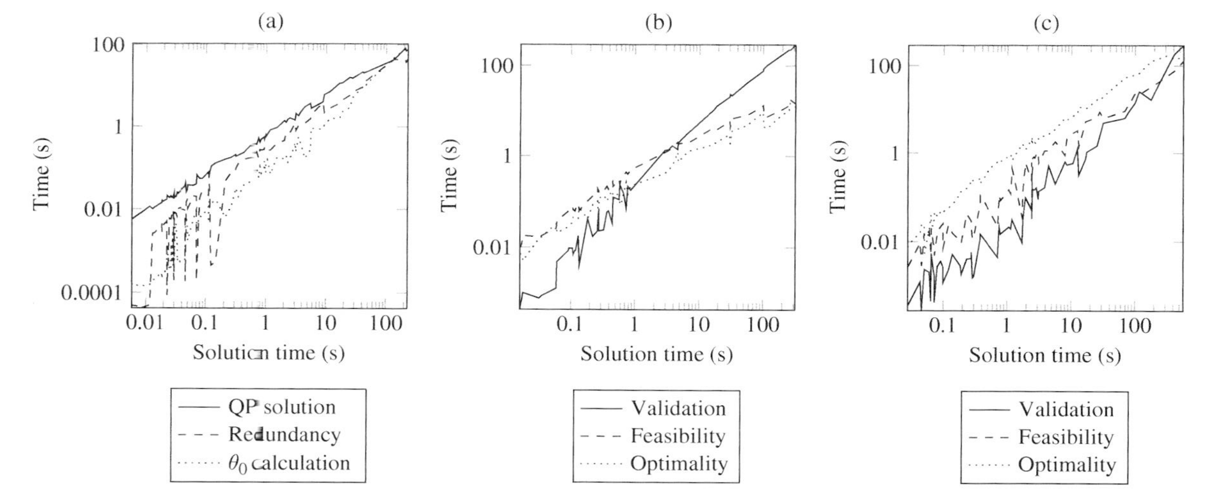

Figure 8.5 The analysis of the computational effort spent on different aspects of the algorithm for the (a) geometrical, (b) combinatorial, and (c) connected-graph algorithm for the test set “POP_mpLP1.”

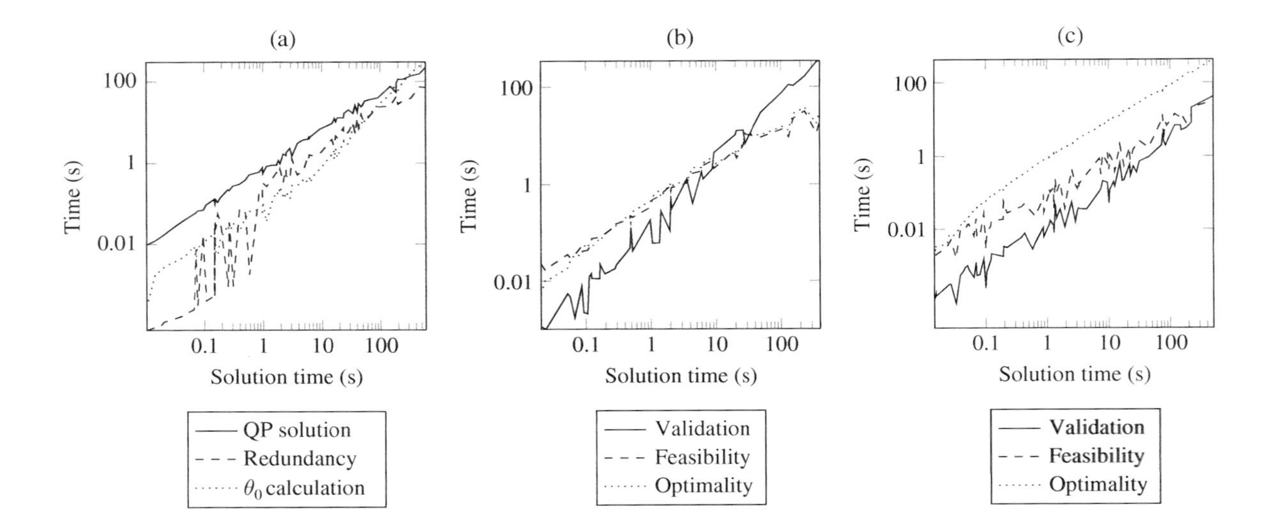

Figure 8.6 The analysis of the computational effort spent on different aspects of the algorithm for the (a) geometrical, (b) combinatorial, and (c) connected-graph algorithm for the test set "POP_mpQP1."

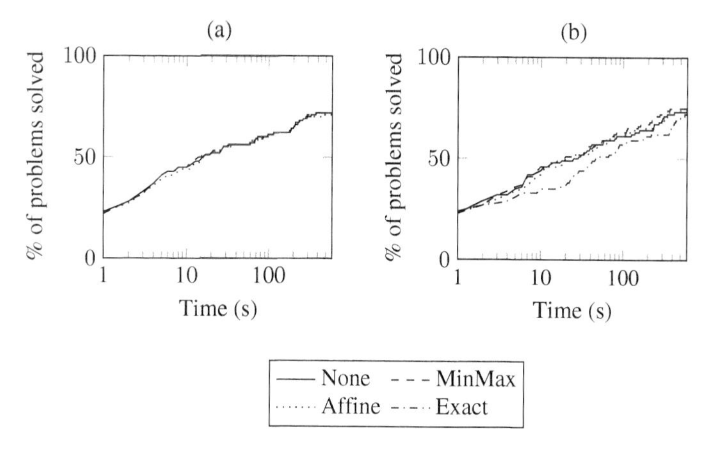

Figure 8.7 The performance of the decomposition algorithm for the different comparison procedures available in POP for the test sets (a) "POP_mpMILP1" and (b) "POP_mpMIQP1."

objective measure regarding the efficiency and stability of various algorithms.

For the continuous case, the computational efficiency of the geometrical and connected-graph algorithms are substantial, especially for the mp-QP case. The key bottlenecks that limit the additional speed-up of these algorithms seem to be the solution of the QP problem and ensuring optimality for the geometrical and connected-graph algorithm, respectively. For the combinatorial approach, the validation procedure seems to be the most demanding, especially for larger scale problems.

For the mixed-integer case, the computational efficiency of using no, a min-max or an affine comparison procedure is very similar. This is due to the fact that the main computational effort is spent in the solution of the mp-QP problem. This is surprising, as Eq. (7.9) is non-convex and thus its solution could be potentially limiting. However, as we used an approximate algorithm without strict error tolerance requirements, this did not cause computational limitations. In addition, it appears that the increased number of partitions resulting from the use of an affine comparison procedure does not impact the computational performance significantly. However, the use of the exact algorithm resulted in an increased

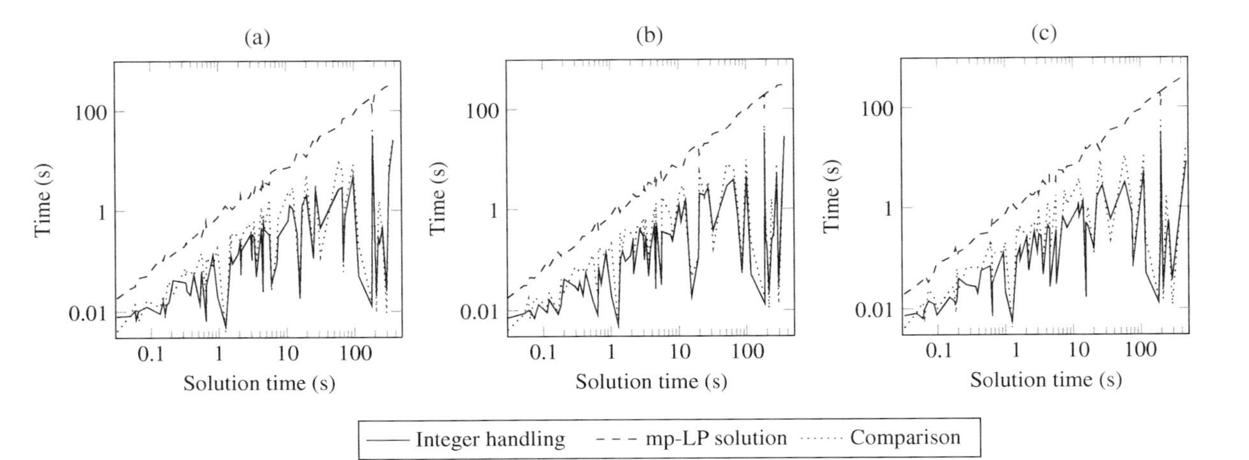

Figure 8.8 The analysis of the computational effort spent on different comparison procedures for the test sets "POP_mpMILP1" for the three different comparison procedures evaluated. Note that due to the linear objective function the affine comparison yields the exact solution, which removes the need to use the "Exact" comparison used for mp-MIQP problems. (a) No comparison. (b) MinMax comparison. (c) Affine comparison.

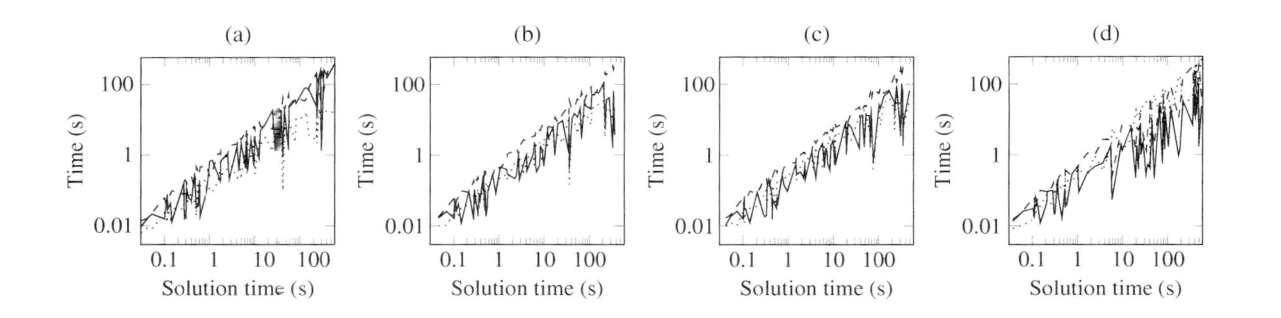

Figure 8.9 The analysis of the computational effort spent on different comparison procedures for the test sets "POP_mpMIQP1" for the four different comparison procedures evaluated. (a) No comparison. (b) MinMax comparison. (c) Affine comparison. (d) Exact comparison.

computational expense. In addition, the calculation of the exact solution for mp-MIQP problems requires the solution of a quadratically constrained feasibility problem. In numerous cases, this led to numerical tolerance issues, as the convergence of the algorithm (the MATLAB in-built `fmincon`) was sometimes not guaranteed.

Acknowledgments

The POP toolbox presented here was released in 2016 by Oberdieck *et al.* [6]. However, this tool has been developed and used by Prof. Pistikopoulos' group since 1999, when the first efforts were made to provide a general-purpose solver for multi-parametric programming. Started predominantly by Nikos Bozinis, POP soon grew more and more to suit the research needs of the group. However, this also meant that revisions were necessary and starting in 2010, Martina Wittmann-Hohlbein led this task and produced the foundation of the POP toolbox as we have it today. Since Prof. Pistikopoulos' move to Texas, Baris Burnak and Justin Katz have also been involved in the development and maintenance of the codebase.

References

1 Kvasnica, M., Grieder, P., Baotić, M., and Morari, M. (2004) Multi-parametric toolbox (MPT), in *Hybrid systems: computation and control, Lecture notes in computer science*, vol. 2993 (eds R. Alur and G.J. Pappas), Springer-Verlag, Berlin, Heidelberg, pp. 448–462, doi: 10.1007/978-3-540-24743-2_30. URL http://dx.doi .org/10.1007/978-3-540-24743-2_30.

2 Herceg, M., Kvasnica, M., Jones, C.N., and Morari, M. (2013) Multi-parametric toolbox 3.0, in *2013 European Control Conference (ECC)*, Zurich, Switzerland (17-19 July 2013), pp. 502–510.

3 Herceg, M., Jones, C.N., Kvasnica, M., and Morari, M. (2015) Enumeration-based approach to solving parametric linear complementarity problems. *Automatica*, 62, 243–248, doi:

10.1016/j.automatica.2015.09.019. URL http://www.sciencedirect.com/science/article/pii/S0005109815003829.

4 Kvasnica, M., Holaza, J., Takacs, B., and Ingole, D. (2015) Design and verification of low-complexity explicit MPC controllers in MPT3, in *2015 European Control Conference (ECC)*, pp. 2595–2600, doi: 10.1109/ECC.2015.7330929.

5 Kvasnica, M., Takács, B., Holaza, J., and Ingole, D. (2015) Reachability analysis and control synthesis for uncertain linear systems in MPT. *IFAC-PapersOnLine*, 48 (14), 302–307, doi: 10.1016/j.ifacol.2015.09.474. URL http://www.sciencedirect.com/science/article/pii/S2405896315015931.

6 Oberdieck, R., Diangelakis, N.A., Papathanasiou, M.M., Nascu, I., and Pistikopoulos, E.N. (2016) POP – parametric optimization toolbox. *Industrial and Engineering Chemistry Research*, 55 (33), 8979–8991, doi: 10.1021/acs.iecr.6b01913. URL http://dx.doi.org/10.1021/acs.iecr.6b01913.

9

Other Developments in Multi-parametric Optimization

In addition to the results on mp-LP, mp-QP, mp-MILP, and mp-MIQP problems presented in the previous chapters, there have also been interesting developments in other areas of multi-parametric programming.

9.1 Multi-parametric Nonlinear Programming

Consider the following nonlinear programming (NLP) problem:

$$\begin{aligned} z = \quad & \underset{x}{\text{Minimize}} \quad f(x) \\ & \text{Subject to} \quad g(x) \leq 0 \\ & \qquad\qquad\quad h(x) = 0 \\ & x \in \mathbb{R}^n, \end{aligned} \tag{9.1}$$

where $f(x) : \mathbb{R}^n \rightarrow \mathbb{R}$, $g(x) : \mathbb{R}^{m \times n} \rightarrow \mathbb{R}^m$ and $h(x) : \mathbb{R}^{s \times n} \rightarrow \mathbb{R}^s$. In many applications some parts of (9.1) might change over time and have great impact on the solution. In order to take this into account, the uncertainty can be considered explicitly by formulating a multi-parametric nonlinear programming (mp-NLP) problem:

$$\begin{aligned} z(\theta) = \quad & \underset{x}{\text{Minimize}} \quad f(x, \theta) \\ & \text{Subject to} \quad g(x, \theta) \leq 0 \\ & \qquad\qquad\quad h(x, \theta) = 0 \\ & \qquad\qquad\quad \theta \in \Theta \subset \mathbb{R}^q, \quad x \in \mathbb{R}^n, \end{aligned} \tag{9.2}$$

Multi-parametric Optimization and Control, First Edition.
Efstratios N. Pistikopoulos, Nikolaos A. Diangelakis, and Richard Oberdieck.

where $f(x) : \mathbb{R}^{(n+q)} \rightarrow \mathbb{R}, g(x) : \mathbb{R}^{m\times(n+q)} \rightarrow \mathbb{R}^m, h(x,\theta) : \mathbb{R}^{s\times(n+q)} \rightarrow \mathbb{R}^s$ and Θ is compact. The difference between problems (9.1) and (9.2) is the presence of the bounded uncertain parameters θ in the constraints. As a result, the solution x, the Lagrange multipliers λ and μ, and the optimal objective function z of problem (9.2) are obtained as a function of θ, i.e. $x(\theta)$, $\lambda(\theta)$, $\mu(\theta)$, and $z(\theta)$, respectively. As these functions may change over the parameter space Θ, the solution to problem (9.2) is given by the partitioning of the feasible parameter space $\Theta_f \subseteq \Theta$, into constrained regions, called critical regions. The strategy for the solution of problems of type (9.2), and for the even more challenging multi-parametric mixed-integer nonlinear programming (mp-MINLP) problems, depends on the convexity of the functions $f(x,\theta)$, $g(x,\theta)$, $h(x,\theta)$, and Θ.

9.1.1 The Convex Case

For the convex case, the first algorithm for the solution of problem (9.2) was presented by Dua and Pistikopoulos [1], which is based on the linearization of $f(x,\theta)$ and $g(x,\theta)$ and the solution of the resulting mp-LP. Due to the convexity of the problem, the maximum error is always attained in one of the vertices. If the error is above a prescribed tolerance ϵ, the parameter space is partitioned and the linearized problem is solved again. This algorithm has been developed further in [2], adapted in [3, 4], and modified in [5, 6] by using convex quadratic approximations for $f(x,\theta)$.

A distinctly different approach to [1] for the solution of problem (9.2) was proposed by Johansen [7]. Instead of approximating $f(x,\theta)$ and $g(x,\theta)$ directly, a deterministic NLP problem is solved at each vertex of a hyper-rectangle $\mathcal{H} \supseteq \Theta$, and the parametric solution is interpolated linearly. If the error between the parametric solution and the exact NLP solution is above a prescribed tolerance ϵ, then $\mathcal{H}$ is subdivided and the procedure is repeated for each hyper-rectangle until the tolerance is met and the algorithm converges. This algorithm has been modified by Bemporad and Filippi to consider simplices instead [8].

Additional approaches consider the use of geometric vertex search [9] and multi-parametric subgradient algorithms [10, 11].

9.1.2 The Non-convex Case

The first algorithm considering non-convex mp-NLP problems was presented by Dua *et al.* [12], where a spatial branch-and-bound approach was used to generate mp-LP problems, which provided parametric lower and upper bounds for the mp-NLP. Additionally, Grancharova and Johansen [13] extended the interpolation algorithm from the convex case to the non-convex case by also allowing for non-linear interpolations of the parametric solution based on the solution of the NLP at the vertices of the hypercube, for which also some heuristics were provided in [14].

However, arguably the most interesting algorithm for polynomial non-convex mp-NLP problems was presented by Fotiou *et al.* [15–18]. In essence, cylindrical algebraic decomposition (CAD) is used to apply recursive projection operations over the parts of the variable space, called cells, which are defined by the so-called projection level factors. This approach was recently extended by Charitopoulos and Dua for the case of multi-parametric mixed integer polynomial programming problems [19]. However, while very exciting from a conceptual standpoint, it is unfortunately limited by the high computational cost of CAD.

Additional approaches consider the use of a moving-front algorithm [20] or the special case of multi-parametric dynamic optimization (mp-DO) [21, 22].

9.2 Dynamic Programming via Multi-parametric Programming

Consider the following multi-stage linear programming problem:

$$
\begin{aligned}
z(x_0) = \quad & \underset{U}{\text{Minimize}} && \sum_{k=0}^{N} c_{x,k}^T x_k + c_{u,k}^T u_k \\
& \text{Subject to} && A_k x_k + S_k u_k \leq b_k, \quad k = 0, \ldots, N \\
& && A_{\text{eq},k} x_k + S_{\text{eq},k} u_k = b_{\text{eq},k}, \quad k = 0, \ldots, N \\
& && x_{k+1} = C_k x_k + D_k u_k + e_k, \quad k = 0, \ldots, N-1 \\
& && X = [x_0^T, x_1^T, \ldots, x_N^T]^T \in \mathbb{R}^{N_x} \\
& && U = [u_0^T, u_1^T, \ldots, u_{N-1}^T]^T \in \mathbb{R}^{N_u},
\end{aligned}
\tag{9.3}
$$

where $c_{x,k} \in \mathbb{R}^{n_x}$, $c_{u,k} \in \mathbb{R}^{n_u}$, $A_k \in \mathbb{R}^{m_k \times n_x}$, $S_k \in \mathbb{R}^{m_k \times n_u}$, $b_k \in \mathbb{R}^{m_k}$, $A_{\text{eq},k} \in \mathbb{R}^{s_k \times n_x}$, $S_{\text{eq},k} \in \mathbb{R}^{s_k \times n_u}$, $b_{\text{eq},k} \in \mathbb{R}^{s_k}$, $C_k \in \mathbb{R}^{v_k \times n_x}$, $D_k \in \mathbb{R}^{v_k \times n_u}$, and $e_k \in \mathbb{R}^{v_k}$.

Remark 9.1 Note that problem (9.3) is a multi-stage LP problem. As shown in the following text, the same strategy can be applied to multi-stage QP, MILP, and MIQP problems.

Dynamic programming is a strategy to solve convex multi-stage optimization problems such as problem (9.3), pioneered predominantly by Bellman in 1950s [23]. It is based on the idea that, starting from the last stage, it is sufficient to solve the optimization problem at each stage individually, and propagate this solution to the previous stage. Therefore, considering problem (9.3), this yields the following dynamic programming problem, which is solved recursively from $k = N$ to $k = 0$:

$$
\begin{aligned}
z_k(x_k) = \quad & \underset{u_k}{\text{Minimize}} \quad c_{x,k}^T x_k + c_{u,k}^T u_k + z_{k+1}(x_{k+1}) \\
& \text{Subject to} \quad A_k x_k + S_k u_k \leq b_k \\
& \qquad\qquad\quad A_{\text{eq},k} x_k + S_{\text{eq},k} u_k = b_{\text{eq},k} \\
& \qquad\qquad\quad x_{k+1} = C_k x_k + D_k u_k + e_k \\
& \qquad\qquad\quad x_k \in \mathbb{R}^{n_x}, \quad u_k \in \mathbb{R}^{n_u}.
\end{aligned}
\tag{9.4}
$$

Classically, problem (9.3) is solved for a specific x_0 by constructing a suitable overestimator for the initial stage $z_N(x_N)$. Conversely, using multi-parametric programming, it is possible to solve problem (9.4) explicitly and thus avoid the use of such an overestimating function. However, this exact solution yields a piecewise affine optimal objective function of the next stage $z_{k+1}(x_{k+1})$. This has two important implications: (i) the optimal objective function of each critical region i, $z_{k+1}^i(x_{k+1})$, needs to be considered individually and (ii) the solutions obtained from considering these different objective functions need to be compared to each other. This has led to two different approaches for mp-DP problems, namely, direct and indirect approaches.

9.2.1 Direct and Indirect Approaches

Within the literature, these issues have been addresses using two different approaches, the direct and the indirect approach:

Direct approach: The initial approach to mp-DP problems, which was presented in 2003 for multi-stage MILP problems [24, 25], was to solve the problem directly by considering each critical region separately and in fact performing the comparison procedure at every stage [26]. In particular for multi-stage MILP and MIQP cases, this direct approach has proven to be computationally interesting, as the comparison procedure would be required in either case.

Indirect approach: While the direct approach is useful for multi-parametric mixed-integer programming problems, the need of the comparison procedure makes its application to multi-stage LP and QP problems computationally unattractive. However, in order to enable the use of mp-DP for multi-stage LP and QP problems, the indirect approach was presented, which avoids the comparison procedure by augmenting the problem formulation with extra parameters [27, 28],[1] an approach that was also successfully applied to multi-stage mp-MILP problems [30, 31]. While the disadvantage of this approach is the increase of the parameter space, it has been successfully applied to a number of problems, even with a moderate number of stages [32–34].

For the special case of mp-DP for multi-stage LP problems, and by extension for each combination of binary variables of a multi-stage MILP problems, Barić *et al.* [35–37] elegantly showed that it is possible to avoid the comparison procedure by utilizing the convexity of the objective function. In particular, they used the fundamental property that any continuous, convex piecewise

---

1 The conceptual idea – lifting the space to avoid the comparison procedure – is equivalent to the work in [29].

function $f(x)$ can be represented using its epigraph [38]:

$$f(x) = \max_i f_i(x) \quad \Leftrightarrow \quad f(x) = \{t \in \mathbb{R} | f_i(x) \leq t, \forall i\}. \tag{9.5}$$

Using Eq. (9.5) in problem (9.4), $z_{k+1}(x_{k+1})$ can be formulated directly. This approach was successfully applied to several case studies [39, 40].

9.3 Multi-parametric Linear Complementarity Problem

Consider the following linear complementarity problem (LCP):

$$\begin{aligned} & w - Mz = q \\ & w \geq 0, \ \ z \geq 0, \ \ w^T z = 0 \\ & w \in \mathbb{R}^n, \ \ z \in \mathbb{R}^n, \end{aligned} \tag{9.6}$$

where $M \in \mathbb{R}^{n \times n}$ and $q \in \mathbb{R}^n$. In order to assess variations in q in problem (9.6), a multi-parametric linear complementarity problem (mp-LCP) is defined as follows:

$$\begin{aligned} & w - Mz = q + Q\theta \\ & w \geq 0, \ \ z \geq 0, \ \ w^T z = 0 \\ & w \in \mathbb{R}^n, \ \ z \in \mathbb{R}^n \\ & \theta \in \Theta = \{\theta \in \mathbb{R}^q | \mathrm{CR}_A \theta \leq \mathrm{CR}_b\}, \end{aligned} \tag{9.7}$$

where $Q \in \mathbb{R}^{n \times q}$ and $x = [w^T, z^T]^T$. The solution to problem (9.7) is given by the partitioning of the feasible parameter space $\Theta_f \subseteq \Theta$ into polytopic regions, called critical regions, each of which is associated with the optimal solution $x(\theta)$, which is an affine function of θ. While in itself an interesting problem, its importance stems from the fact that the parametric Karush–Kuhn–Tucker conditions for mp-LP and mp-QP problems are of the form of problem (9.7) (see Chapter 3.1.2). Thus, an algorithm to solve problem (9.7) directly solves mp-LP and mp-QP problems as well.

Similarly to mp-LP and mp-QP problems, most solution approaches for mp-LCP problems can be subdivided into two categories: geometrical [41–43] and combinatorial [44] approaches. In [41], given the basis of a full dimensional region CR, the set of bases that are adjacent to CR are calculated, similarly to [45]. This result is refined in [42], which yields an output-sensitive algorithm and in [43], which provides an efficient two-stage procedure for the solution of (9.7). Conversely, [44] employs the algorithm from [46] for the identification of all feasible and optimal active sets. Note that the case of multi-parametric mixed-integer linear complementarity problems is considered via the exhaustive enumeration of all combinations.

A distinctly different approach was presented by Li and Ierapetritou [47], where the problem is reformulated into a mixed-integer linear complementarity problem, which is subsequently solved. Note that this paper also considers multi-parametric mixed-integer linear complementarity problems by solving an MILP problem to identify a candidate combination of binary variables.

Remark 9.2 In addition, Kalashnikov *et al.* [48] recently provided several sufficient conditions that guarantee monotonicity of the solutions of an mp-LCP problem.

9.4 Inverse Multi-parametric Programming

So far, the solution of the multi-parametric programming problems has been considered. However, in 2008 independently of each other, Jones *et al.* [49] and Baes *et al.* [50] considered the opposite question: given the solution of a multi-parametric programming problem, is it possible to reconstruct the original problem formulation? This type of problem has been named inverse multi-parametric programming, and it has been considered by several researchers due to its theoretical and practical properties [51–61]. In particular, the following statements have been proven to hold:

- Every continuous nonlinear control system can be obtained by parametric convex programming [50].
- Every continuous piecewise affine function can be obtained by solving a parametric linear program y[52].

- Any discontinuous piecewise affine function is the optimal solution to a parametric linear programming problem [58].
- Inverse multi-parametric programming problems can be solved via convex liftings [54, 56].

9.5 Bilevel Programming Using Multi-parametric Programming

Optimization problems involving two decision makers at two different decision levels are referred to as bilevel programming problems: the first decision maker (upper level) is solving an optimization problem, which includes in its constraint set another optimization problem solved by the second decision maker (lower level), i.e.

$$
\begin{aligned}
\underset{x,y}{\text{Minimize}} \quad & F(x,y) \\
\text{Subject to} \quad & G(x,y) \leq 0 \\
& H(x,y) = 0 \\
& \underset{y}{\text{Minimize}} \quad f(x,y) \\
& \text{Subject to} \quad g(x,y) \leq 0 \\
& \qquad\qquad\quad h(x,y) = 0,
\end{aligned}
\tag{9.8}
$$

where $x \in \mathbb{R}^{n_1}$ and $y \in \mathbb{R}^{n_2}$, $F(x,y) : \mathbb{R}^{(n_1+n_2)} \to \mathbb{R}$, $G(x,y) : \mathbb{R}^{m_1 \times (n_1+_2)}$, $H(x,y) : \mathbb{R}^{s_1 \times (n_1+_2)}$, $f(x,y) : \mathbb{R}^{(n_1+n_2)} \to \mathbb{R}$, $g(x,y) : \mathbb{R}^{m_2 \times (n_1+_2)}$, and $h(x,y) : \mathbb{R}^{s_2 \times (n_1+_2)}$. Bilevel programming problems are very challenging to solve, even in the linear case [62, 63], and thus classical approaches have mainly be presented for problems containing only continuous variables. Conversely, the use of multi-parametric programming enables the solution of the lower level problem as a function of the decision variables of the upper level problem. The resulting exact parametric solutions are then substituted into the upper level problem, which can be solved as a set of single-level deterministic programming problems. This strategy has been applied to a variety of combinations of LP, QP, MILP, and MINLP problems [64–69].

Remark 9.3 In order to consider mixed-integer problems on the upper level, it is necessary to consider binary variables as parameters to the lower level problem. This issue can be addressed by either performing an exhaustive enumeration over all possible combinations of binary variables or using a recently proposed framework for the handling of binary parameters [70].

9.6 Multi-parametric Multi-objective Optimization

Consider the following multi-objective optimization (MOO) problem:

$$
\begin{aligned}
\underset{x}{\text{Minimize}} \quad & \{f_1(x), f_2(x), \dots, f_N(x)\} \\
\text{Subject to} \quad & Ax \leq b \\
& x \in \mathbb{R}^n,
\end{aligned}
\tag{9.9}
$$

where $A \in \mathbb{R}^{m \times n}$ and $b \in \mathbb{R}^m$. The solution to problems of type (9.9) is given by so-called Pareto points, i.e. a point x^* where there does not exist a point $\hat{x}$ such that there exists $f_i(\hat{x}) < f_i(x^*)$ and $f_j(\hat{x}) \leq f_j(x^*)$, $j \neq i$. The set of all Pareto points is called the Pareto front. The two most common strategies to obtain a Pareto point is (a) the ϵ-constraint method,[2] i.e.

$$
\begin{aligned}
\underset{x}{\text{Minimize}} \quad & f_1(x) \\
\text{Subject to} \quad & f_j(x) \leq \epsilon_j, \quad \forall j = 2, \dots, N \\
& Ax \leq b \\
& x \in \mathbb{R}^n,
\end{aligned}
\tag{9.10}
$$

where the parameter ϵ_j denotes an upper bound on the function $f_j(x)$, and (b) the linear scalarization method,[3] i.e.

$$
\begin{aligned}
\underset{x}{\text{Minimize}} \quad & \sum_{i=1}^{N} w_i f_i(x) \\
\text{Subject to} \quad & Ax \leq b \\
& w_i \geq 0, \forall i = 1, \dots, N \\
& \sum_{i=1}^{k} w_i = 1 \\
& x \in \mathbb{R}^n.
\end{aligned}
\tag{9.11}
$$

2 Note that the choice of $f_1(x)$ as the objective function is arbitrary (see [71]).
3 This method is sometimes also referred to as weighting method [71].

However, in both strategies the solution of the MOO problem depends on the values of certain parameters, namely, ϵ_j and w_i. Hence, while many researchers consider the iterative solution of the resulting optimization problems for different parameter values [72, 73], some attention has been given to the explicit calculation of the entire Pareto front via multi-parametric programming, which solves optimization problems as a function and for a range of certain parameters. In [74, 75] the authors consider the case of linear cost functions, and in [76] the case of a mixed-integer nonlinear MOO was considered. The case of quadratic cost functions was treated in [77, 78], although either only conceptually or for the case where the quadratic part remains constant. Recently, approaches for general convex quadratic cost functions [79] as well as uncertain mp-MOO problems with linear objective functions [80] were considered.

Remark 9.4 Note that in [81], the question of multi-objective model predictive control via multi-parametric programming was considered.

References

1 Dua, V. and Pistikopoulos, E.N. (1999) Algorithms for the solution of multiparametric mixed-integer nonlinear optimization problems. *Industrial and Engineering Chemistry Research*, 38 (10), 3976–3987, doi: 10.1021/ie980792u.
2 Acevedo, J. and Salgueiro, M. (2003) An efficient algorithm for convex multiparametric nonlinear programming problems. *Industrial and Engineering Chemistry Research*, 42 (23), 5883–5890, doi: 10.1021/ie0301278. URL http://dx.doi.org/10.1021/ie0301278.
3 Wittmann-Hohlbein, M. and Pistikopoulos, E.N. (2012) A two-stage method for the approximate solution of general multiparametric mixed-integer linear programming problems. *Industrial and Engineering Chemistry Research*, 51 (23), 8095–8107, doi: 10.1021/ie201408p.

4 Wittmann-Hohlbein, M. and Pistikopoulos, E.N. (2013) On the global solution of multi-parametric mixed integer linear programming problems. *Journal of Global Optimization*, 57 (1), 51–73, doi: 10.1007/s10898-012-9895-2. URL http://dx.doi.org/10.1007/s10898-012-9895-2.

5 Johansen, T.A. (2002) On multi-parametric nonlinear programming and explicit nonlinear model predictive control, in *Proceedings of the 41st IEEE Conference on Decision and Control, 2002*, vol. 3, pp. 2768–2773, doi: 10.1109/CDC.2002.1184260.

6 Domínguez, L.F. and Pistikopoulos, E.N. (2013) A quadratic approximation-based algorithm for the solution of multiparametric mixed-integer nonlinear programming problems. *AIChE Journal*, 59 (2), 483–495, doi: 10.1002/aic.13838. URL http://dx.doi.org/10.1002/aic.13838.

7 Johansen, T.A. (2004) Approximate explicit receding horizon control of constrained nonlinear systems. *Automatica*, 40 (2), 293–300, doi: 10.1016/j.automatica.2003.09.021. URL http://www.sciencedirect.com/science/article/pii/S0005109803003431.

8 Bemporad, A. and Filippi, C. (2006) An algorithm for approximate multiparametric convex programming. *Computational Optimization and Applications*, 35 (1), 87–108, doi: 10.1007/s10589-006-6447-z. URL http://dx.doi.org/10.1007/s10589-006-6447-z.

9 Narciso, D.A. (2009) *Developments in nonlinear multiparametric programming and control*, Ph.D. thesis, Imperial College, London.

10 Leverenz, J. (2015) Network target coordination for multiparametric programming, Ph.D. thesis, Clemson University, Clemson, SC. URL http://search.proquest.com/docview/1727773829.

11 Leverenz, J., Xu, M., and Wiecek, M.M. (2016) Multiparametric optimization for multidisciplinary engineering design. *Structural and Multidisciplinary Optimization*, 54 (4), 1–16, doi: 10.1007/s00158-016-1437-y. URL http://dx.doi.org/10.1007/s00158-016-1437-y.

12 Dua, V., Papalexandri, K.P., and Pistikopoulos, E.N. (2004) Global optimization issues in multipara-

metric continuous and mixed-integer optimization problems. *Journal of Global Optimization*, 30 (1), 59–89, doi: 10.1023/B:JOGO.0000049091.73047.7e. URL http://dx.doi.org/10.1023/B%3AJOGO.0000049091.73047.7e.

13 Grancharova, A. and Johansen, T.A. (2006) Explicit approximate approach to feedback min-max model predictive control of constrained nonlinear systems, in *Proceedings of the 45th IEEE Conference on Decision and Control*, pp. 4848–4853, doi: 10.1109/CDC.2006.377772.

14 Grancharova, A., Johansen, T.A., and Tøndel, P. (2007) Computational aspects of approximate explicit nonlinear model predictive control, in *Assessment and future directions of nonlinear model predictive control* (eds R. Findeisen, F. Allgöwer, and L.T. Biegler), Springer-Verlag, Berlin, Heidelberg, pp. 181–192, doi: 10.1007/978-3-540-72699-9.14. URL http://dx.doi.org/10.1007/978-3-540-72699-9.14.

15 Fotiou, I.A., Parrilo, P.A., and Morari, M. (2005) Nonlinear parametric optimization using cylindrical algebraic decomposition, in *Proceedings of the 44th IEEE Conference on Decision and Control, 2005 and 2005 European Control Conference. CDC-ECC '05*, pp. 3735–3740, doi: 10.1109/CDC.2005.1582743.

16 Fotiou, I.A., Beccuti, A.G., Papafotiou, G., and Morari, M. (2006) Optimal control of piece-wise polynomial hybrid systems using cylindrical algebraic decomposition, in *HSCC (Hybrid Systems: Computation and Control)*, Santa Barbara, CA, pp. 227–241. URL http://control.ee.ethz.ch/index.cgi?page=publications;action=details;id=2335.

17 Fotiou, I.A., Rostalski, P., Parrilo, P.A., and Morari, M. (2006) Parametric optimization and optimal control using algebraic geometry methods. *International Journal of Control*, 79 (11), 1340–1358. URL http://control.ee.ethz.ch/index.cgi?page=publications;action=details;id=2387.

18 Fotiou, I.A., Rostalski, P., Sturmfels, B., and Morari, M. (2006) An algebraic geometry approach to nonlinear parametric optimization in control, in *American Control Conference*, Minneapolis, MN, pp. 3618–3623. URL http://control.ee.ethz.ch/index.cgi?page=publications;action=details;id=2224.

19 Charitopoulos, V.M. and Dua, V. (2016) Explicit model predictive control of hybrid systems and multiparametric mixed integer polynomial programming. *AIChE Journal*, 62 (9), 3441–3460, doi: 10.1002/aic.15396. URL http://dx.doi.org/10.1002/aic.15396.

20 Hale, E.T. (2005) *Numerical methods for d-parametric nonlinear programming with chemical process control and optimization applications*, Ph.D. thesis, University of Austin, Austin, TX, USA. URL http://www.lib.utexas.edu/etd/d/2005/halee37108/halee37108.pdf.

21 Sakizlis, V., Perkins, J.D., and Pistikopoulos, E.N. (2005) Explicit solutions to optimal control problems for constrained continuous-time linear systems. *IEEE Proceedings: Control Theory and Applications*, 152 (4), 443–452, doi: 10.1049/ip-cta:20059041.

22 Sun, M., Chachuat, B., and Pistikopoulos, E.N. (2016) Design of multi-parametric NCO tracking controllers for linear dynamic systems. *Computers and Chemical Engineering*, 92, 64–77, doi: 10.1016/j.compchemeng.2016.04.038. URL http://www.sciencedirect.com/science/article/pii/S009813541630134X.

23 Bellman, R. (2003) *Dynamic programming*, Dover Publications, Mineola, NY, dover ed edn.

24 Borrelli, F., Baotic, M., Bemporad, A., and Morari, M. (2003) An efficient algorithm for computing the state feedback optimal control law for discrete time hybrid systems, in *Proceedings of the 2003 American Control Conference, 2003*, vol. 6, pp. 4717–4722, doi: 10.1109/ACC.2003.1242468.

25 Baotić, M., Christophersen, F.J., and Morari, M. (2003) A new algorithm for constrained finite time optimal control of hybrid systems with a linear performance index, in *Proceedings of the European Control Conference*, Cambridge, UK.

26 Borrelli, F., Baotić, M., Bemporad, A., and Morari, M. (2005) Dynamic programming for constrained optimal control of discrete-time linear hybrid systems. *Automatica*, 41 (10), 1709–1721, doi: 10.1016/j.automatica.2005.04.017. URL http://www.sciencedirect.com/science/article/pii/S0005109805001524.

27 Faísca, N.P., Kouramas, K.I., Saraiva, P.M., Rustem, B., and Pistikopoulos, E.N. (2007) Robust dynamic programming via multi-parametric programming, in *17th European Symposium on Computer Aided Process Engineering, Computer aided chemical engineering*, vol. 24 (eds V. Pleşu and P. Şerban Agachi), Elsevier, pp. 811–816, doi: 10.1016/S1570-7946(07)80158-1. URL http://www.sciencedirect.com/science/article/pii/S1570794607801581.

28 Faísca, N.P., Kouramas, K.I., Saraiva, P.M., Rustem, B., and Pistikopoulos, E.N. (2008) A multi-parametric programming approach for constrained dynamic programming problems. *Optimization Letters*, 2 (2), 267–280, doi: 10.1007/s11590-007-0056-3. URL http://dx.doi.org/10.1007/s11590-007-0056-3.

29 Fuchs, A., Axehill, D., and Morari, M. (2015) Lifted evaluation of mp-MIQP solutions. *IEEE Transactions on Automatic Control*, 60 (12), 3328–3331, doi: 10.1109/TAC.2015.2417853.

30 Rivotti, P. and Pistikopoulos, E.N. (2014) Constrained dynamic programming of mixed-integer linear problems by multi-parametric programming. *Computers and Chemical Engineering*, 70, 172–179, doi: 10.1016/j.compchemeng.2014.03.021. URL http://www.sciencedirect.com/science/article/pii/S0098135414001021.

31 Rivotti, P. and Pistikopoulos, E.N. (2014) A dynamic programming based approach for explicit model predictive control of hybrid systems. *Computers and Chemical Engineering*, 72, 126–144, doi: 10.1016/j.compchemeng.2014.06.003. URL http://www.sciencedirect.com/science/article/pii/S0098135414001823.

32 Kouramas, K.I., Faísca, N.P., Panos, C., and Pistikopoulos, E.N. (2011) Explicit/multi-parametric model predictive control (MPC) of linear discrete-time systems by dynamic and multi-parametric programming. *Automatica*, 47 (8), 1638–1645, doi: 10.1016/j.automatica.2011.05.001. URL http://www.sciencedirect.com/science/article/pii/S000510981100255X.

33 Kouramas, K.I., Panos, C., and Pistikopoulos, E.N. (2011) Algorithm for robust explicit/multi-parametric MPC in embedded control systems. *IFAC Proceedings Volumes (IFAC-PapersOnline)*, 18, 1344–1349, doi: 10.3182/20110828-6-IT-1002.02126. URL http://dx.doi.org/10.3182/20110828-6-IT-1002.02126.

34 Kouramas, K.I., Panos, C., Faísca, N.P., and Pistikopoulos, E.N. (2013) An algorithm for robust explicit/multi-parametric model predictive control. *Automatica*, 49 (2), 381–389, doi: 10.1016/j.automatica.2012.11.035. URL http://www.sciencedirect.com/science/article/pii/S0005109812005717.

35 Baric, M., Grieder, P., Baotic, M., and Morari, M. (2005) Optimal control of PWA systems by exploiting problem structure, in *IFAC World Congress*, Prague, Czech Republic. URL http://control.ee.ethz.ch/index.cgi?action=details;id=2033;page=publications.

36 Barić, M., Grieder, P., Baotić, M., and Morari, M. (2008) An efficient algorithm for optimal control of PWA systems with polyhedral performance indices. *Automatica*, 44 (1), 296–301, doi: 10.1016/j.automatica.2007.05.005. URL http://www.sciencedirect.com/science/article/pii/S0005109807002452.

37 Baric, M., Rakovic, S.V., Besselmann, T., and Morari, M. (2008) Max-min optimal control of constrained discrete-time systems, in *IFAC World Congress*, Seoul, Korea, pp. 8803–8808. URL https://control.ee.ethz.ch/index.cgi?page=publications;action=details;id=3031.

38 Boyd, S.P. and Vandenberghe, L. (2004) *Convex optimization*, Cambridge University Press, Cambridge, UK and New York.

39 Besselmann, T., Lofberg, J., and Morari, M. (2008) Explicit model predictive control for linear parameter-varying systems, in *47th IEEE Conference on Decision and Control, 2008. CDC 2008*, pp. 3848–3853, doi: 10.1109/CDC.2008.4738798.

40 Besselmann, T., Lofberg, J., and Morari, M. (2012) Explicit MPC for LPV systems: stability and optimality. *IEEE Transactions on Automatic Control*, 57 (9), 2322–2332, doi: 10.1109/TAC.2012.2187400.

41 Jones, C.N. and Morari, M. (2006) Multiparametric linear complementarity problems, in *2006 45th IEEE Conference on Decision and Control*, pp. 5687–5692, doi: 10.1109/CDC.2006.377797.

42 Columbano, S., Fukuda, K., and Jones, C.N. (2009) An output-sensitive algorithm for multi-parametric LCPs with sufficient matrices, in *CRM Proceedings and Lecture Notes*, vol. 48, 1–30, doi: 10.1090/crmp/048/04. URL http://control.ee.ethz.ch/index.cgi?page=publications;action=details;id=3196.

43 Adelgren, N. and Wiecek, M.M. (2016) A two-phase algorithm for the multiparametric linear complementarity problem. *European Journal of Operational Research*, 254 (3), 715–738, doi: 10.1016/j.ejor.2016.04.043. URL http://www.sciencedirect.com/science/article/pii/S0377221716302806.

44 Herceg, M., Jones, C.N., Kvasnica, M., and Morari, M. (2015) Enumeration-based approach to solving parametric linear complementarity problems. *Automatica*, 62, 243–248, doi: 10.1016/j.automatica.2015.09.019. URL http://www.sciencedirect.com/science/article/pii/S0005109815003829.

45 Tøndel, P., Johansen, T.A., and Bemporad, A. (2003) An algorithm for multi-parametric quadratic programming and explicit MPC solutions. *Automatica*, 39 (3), 489–497, doi: 10.1016/S0005-1098(02)00250-9. URL http://www.sciencedirect.com/science/article/pii/S0005109802002509.

46 Gupta, A., Bhartiya, S., and Nataraj, P. (2011) A novel approach to multiparametric quadratic programming. *Automatica*, 47 (9), 2112–2117, doi: 10.1016/j.automatica.2011.06.019. URL http://www.sciencedirect.com/science/article/pii/S0005109811003190.

47 Li, Z. and Ierapetritou, M.G. (2010) A method for solving the general parametric linear complementarity problem. *Annals of Operations Research*, 181 (1), 485–501, doi: 10.1007/s10479-010-0770-6. URL http://dx.doi.org/10.1007/s10479-010-0770-6.

48 Kalashnikov, V.V., Kalashnykova, N.I., and Castillo-Pérez, F.J. (2016) Solutions of parametric complementarity problems monotone with respect to parameters. *Journal of Global Optimization*, 64 (4), 703–719, doi: 10.1007/s10898-015-0369-1. URL http://dx.doi.org/10.1007/s10898-015-0369-1.

49 Jones, C.N., Kerrigan, E.C., and Maciejowski, J.M. (2008) On polyhedral projection and parametric programming. *Journal of Optimization Theory and Applications*, 138, 207–220.

50 Baes, M., Diehl, M., and Necoara, I. (2008) Every continuous nonlinear control system can be obtained by parametric convex programming. *IEEE Transactions on Automatic Control*, 53 (8), 1963–1967, doi: 10.1109/TAC.2008.928131.

51 Hempel, A.B., Goulart, P.J., and Lygeros, J. (2012) Inverse parametric quadratic programming and an application to hybrid control, in *Nonlinear Model Predictive Control*, Noordwijkerhout, NL, pp. 68–73. URL http://control.ee.ethz.ch/index.cgi?page=publications;action=details;id=4074.

52 Hempel, A.B., Goulart, P.J., and Lygeros, J. (2013) Every continuous piecewise affine function can be obtained by solving a parametric linear program, in *2013 European Control Conference (ECC),Zurich, Switzerland (17–19 July 2013)*, pp. 2657–2662.

53 Nguyen, N.A., Olaru, S., Rodriguez-Ayerbe, P., Hovd, M., and Necoara, I. (2014) Inverse parametric convex programming problems Via convex liftings, in *World Congress*, IFAC, Elsevier, IFAC proceedings volumes, pp. 2489–2494, doi: 10.3182/20140824-6-ZA-1003.02364.

54 Nguyen, N.A., Olaru, S., Rodriguez-Ayerbe, P., Hovd, M., and Necoara, I. (2014) On the lifting problems and their connections with piecewise affine control law design, in *2014 European Control Conference (ECC)*, pp. 2164–2169, doi: 10.1109/ECC.2014.6862605.

55 Nguyen, N.A., Olaru, S., Rodriguez-Ayerbe, P., Hovd, M., and Necoara, I. (2015) Fully inverse parametric linear/quadratic programming problems via convex liftings, in *Developments in model-based optimization and control, Lecture notes in control and information sciences*, vol. 464 (eds S. Olaru, A. Grancharova, and F. Lobo Pereira), Springer International Publishing, pp. 27–47, doi: 10.1007/978-3-319-26687-9.2. URL http://dx.doi.org/10.1007/978-3-319-26687-9.2.

56 Nguyen, N.A., Olaru, S., and Rodriguez-Ayerbe, P. (2015) On the complexity of the convex liftings-based solution to inverse parametric convex programming problems, in *2015 European Control Conference (ECC)*, pp. 3428–3433, doi: 10.1109/ECC.2015.7331064.

57 Nguyen, N.A., Rodriguez-Ayerbe, P., and Olaru, S. (2015) Inverse parametric linear/quadratic programming problem for continuous PWA functions defined on polyhedral partitions of polyhedra, in *IEEE Conference on Decision and Control (CDC)*, OSAKA, Japan, pp. 5920–5925.

58 Nguyen, N.A., Rodriguez-Ayerbe, P., and Olaru, S. (2015) Any discontinuous PWA function is optimal solution to a parametric linear programming problem, in *IEEE Conference on Decision and Control (CDC)*, OSAKA, Japan, pp. 5926–5931.

59 Nguyen, N.A., Olaru, S., Rodriguez-Ayerbe, P., Hovd, M., and Ion, N. (2015) *Constructive solution of inverse parametric linear/quadratic programming problems.* HAL Archives-ouvertes, hal-01207234v2.

60 Gulan, M., Nguyen, N.A., Olaru, S., Rodriguez-Ayerbe, P., and Rohal'-Ilkiv, B. (2015) Implications of inverse parametric optimization in model predictive control, in (eds. S. Olaru, A. Grancharova, F. Lobo Pereira)*Developments in model-based optimization and control, Distributed control and industrial applications*, Springer International Publishing, Cham, pp. 49–70, doi: 10.1007/978-3-319-26687-9.3. URL http://dx.doi.org/10.1007/978-3-319-26687-9.3.

61 Hempel, A.B., Goulart, P.J., and Lygeros, J. (2015) Inverse parametric optimization with an application to hybrid system control. *IEEE Transactions on Automatic Control*, 60 (4), 1064–1069. URL http://ieeexplore.ieee.org/stamp/stamp.jsp?arnumber=6849937.

62 Hansen, P., Jaumard, B., and Savard, G. (1992) New branch-and-bound rules for linear bilevel programming. *SIAM Journal on Scientific and Statistical Computing*, 13 (5), 1194–1217.

63 Vicente, L., Savard, G., and Judice, J. (1994) Descent approaches for quadratic bilevel programming. *Journal of Optimization Theory and Applications*, 81 (2), 379–399.

64 Pistikopoulos, E.N., Dua, V., and Ryu, J.H. (2003) Global optimization of bilevel programming problems via parametric programming. *Frontiers in Global Optimization*, 74, 457–476.

65 Ryu, J.H., Dua, V., and Pistikopoulos, E.N. (2004) A bilevel programming framework for enterprise-wide process networks under uncertainty. *Computers and Chemical Engineering*, 28 (6–7), 1121–1129.

66 Faisca, N.P., Dua, V., Saraiva, P.M., Rustem, B., and Pistikopoulos, E.N. (2006) A global parametric programming optimisation strategy for multilevel problems, in *16th European Symposium on Computer Aided Process Engineering and*

9th International Symposium on Process Systems Engineering, Garmisch-Partenkirchen, Germany (9–13 July 2006), vol. 21, pp. 215–220.

67 Faísca, N.P., Dua, V., Rustem, B., Saraiva, P.M., and Pistikopoulos, E.N. (2007) Parametric global optimisation for bilevel programming. *Journal of Global Optimization*, 38 (4), 609–623, doi: 10.1007/s10898-006-9100-6. URL http://dx.doi.org/10.1007/s10898-006-9100-6.

68 Faísca, N., Saraiva, P., Rustem, B., and Pistikopoulos, E. (2009) A multi-parametric programming approach for multilevel hierarchical and decentralised optimisation problems. *Computational Management Science*, 6 (4), 377–397, doi: 10.1007/s10287-007-0062-z. URL http://dx.doi.org/10.1007/s10287-007-0062-z.

69 Domínguez, L.F. and Pistikopoulos, E.N. (2010) Multiparametric programming based algorithms for pure integer and mixed-integer bilevel programming problems. *Computers and Chemical Engineering*, 34 (12), 2097–2106, doi: 10.1016/j.compchemeng.2010.07.032. URL http://www.sciencedirect.com/science/article/pii/S0098135410002802.

70 Oberdieck, R., Diangelakis, N.A., Avraamidou, S., and Pistikopoulos, E.N. (2017) On unbounded and binary parameters in multi-parametric programming: applications to mixed-integer bilevel optimization and duality theory. *Journal of Global Optimization*, 69 (3), 587–606, available online. URL http://link.springer.com/article/10.1007/s10898-016-0463-z.

71 Miettinen, K. (1998) *Nonlinear multiobjective optimization, International series in operations research & management science*, vol. 12, Springer US, Boston, MA.

72 Capitanescu, F., Ahmadi, A., Benetto, E., Marvuglia, A., and Tiruta-Barna, L. (2015) Some efficient approaches for multi-objective constrained optimization of computationally expensive black-box model problems. *Computers and Chemical Engineering*, 82, 228–239, doi: 10.1016/j.compchemeng.2015.07.013. URL http://www.sciencedirect.com/science/article/pii/S0098135415002483.

73 Antipova, E., Pozo, C., Guillén-Gosálbez, G., Boer, D., Cabeza, L.F., and Jiménez, L. (2015) On the use of filters to facilitate the post-optimal analysis of the Pareto solutions in multi-objective optimization. *Computers and Chemical Engineering*, 74, 48–58, doi: 10.1016/j.compchemeng.2014.12.012. URL http://www.sciencedirect.com/science/article/pii/S009813541400341X.

74 Gass, S. and Saaty, T. (1955) The computational algorithm for the parametric objective function. *Naval Research Logistics Quarterly*, 2 (1–2), 39–45, doi: 10.1002/nav.3800020106. URL http://dx.doi.org/10.1002/nav.3800020106.

75 Yuf, P.L. and Zeleny, M. (1976) Linear multiparametric programming by multicriteria simplex method. *Management Science*, 23 (2), 159–170, doi: 10.1287/mnsc.23.2.159. URL http://dx.doi.org/10.1287/mnsc.23.2.159.

76 Papalexandri, K.P. and Dimkou, T.I. (1998) A parametric mixed-integer optimization algorithm for multiobjective engineering problems involving discrete decisions. *Industrial and Engineering Chemistry Research*, 37 (5), 1866–1882, doi: 10.1021/ie970720n.

77 Goh, C.J. and Yang, X.Q. (1996) Analytic efficient solution set for multi-criteria quadratic programs. *European Journal of Operational Research*, 92 (1), 166–181, doi: 10.1016/0377-2217(95)00040-2. URL http://www.sciencedirect.com/science/article/pii/0377221795000402.

78 Ghaffari-Hadigheh, A., Romanko, O., and Terlaky, T. (2010) Bi-parametric convex quadratic optimization. *Optimization Methods and Software*, 25 (2), 229–245, doi: 10.1080/10556780903239568. URL http://dx.doi.org/10.1080/10556780903239568.

79 Oberdieck, R. and Pistikopoulos, E.N. (2016) Multi-objective optimization with convex quadratic cost functions: a multi-parametric programming approach. *Computers and Chemical Engineering*, 85, 36–39, doi: 10.1016/j.compchemeng.2015.10.011. URL http://www.sciencedirect.com/science/article/pii/S0098135415003269.

80 Charitopoulos, V.M. and Dua, V. (2016) A unified framework for model-based multi-objective linear process and energy optimisation under uncertainty. *Applied Energy*,

186, doi: 10.1016/j.apenergy.2016.05.082. URL http://www.sciencedirect.com/science/article/pii/S0306261916306791.

81 Bemporad, A. and Munoz de la Pena, D. (2009) Multiobjective model predictive control based on convex piecewise affine costs, in *2009 European Control Conference (ECC),Budapest, Hungary (23–26 August 2009)*, pp. 2402–2407.

Part II

Multi-parametric Model Predictive Control

10

Multi-parametric/Explicit Model Predictive Control

10.1 Introduction

Model predictive control (MPC) has become the accepted standard for complex constrained multivariable control problems in the process industries [1]. Starting from the current state of the process, an open-loop optimal control problem is solved over a finite time horizon. The computation is repeated at the next time step, considering the new process state and over a shifted horizon. This approach is typically referred to as a rolling/receding horizon policy. The concept of the receding horizon policy is presented in Figure 10.1 and briefly discussed in the following text.

In Figure 10.1a,b, the system/process goal is to achieve a certain operating objective denoted by the dash-dotted line, i.e. the trajectory of the (predicted) system outputs to match this of the set-point. This is achieved by the manipulation of input variables, which play the role of degrees of freedom and will be eventually treated as optimization variables. The dotted line in both figures denotes the present time. Continuous lines denote actions and behaviors that have already happened and dashed lines denote actions to be taken into the future and predicted behaviors. In MPC, the prediction of the future actions, based on a linear/non-linear discrete-time model, takes place over finite time horizons.[1] Those horizons can

1 The interested reader is referred to [2–4] for a complete overview of the theory and application practices of MPC.

Multi-parametric Optimization and Control, First Edition.
Efstratios N. Pistikopoulos, Nikolaos A. Diangelakis, and Richard Oberdieck.

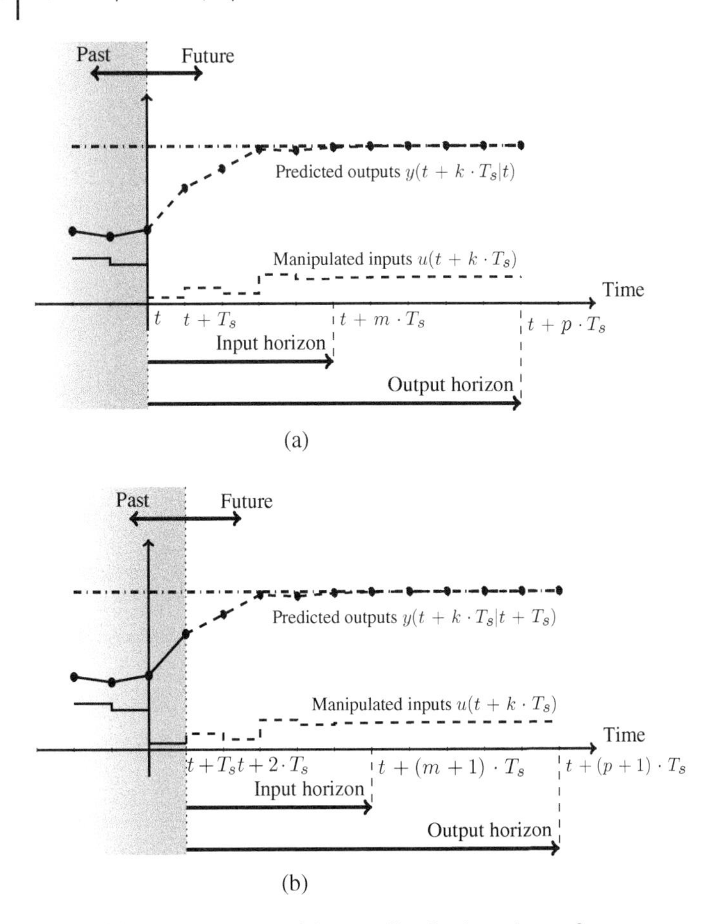

Figure 10.1 The concept of the receding horizon. A set of m manipulated inputs is calculated. In (a) the problem is solved at time t. The system propagates for T_s and re-solved in (b) at time $t + T_s$. The move of the "present time front" is shown with the dotted line and gray area to the left.

be categorized as input (or control) and output horizons. The input horizon, always smaller than or equal to the output horizon, is the number of time steps (T_s) for which a control action is calculated. The output horizon is the time steps for which the behavior of the system is predicted. In Figure 10.1 they are denoted by the letters m and p, respectively. Note that in the cases where $m < p$ then for control actions in the time period $m + 1, \ldots, p$ a rule is considered, typically equality of those actions to the action at time step m. At present time t, the optimization problem formulation corresponding to the MPC is solved and the vector of future inputs is computed (Figure 10.1a). The input action corresponding to the first time step is only applied thus causing the system to propagate in time by T_s. The procedure is repeated at $t + T_s$ over input and output horizons of m and p, respectively. This moving horizon policy respects all input and output constraints, optimizes a performance index, and is applied recursively.

In this chapter, we discuss multi-parametric/explicit MPC (mp-MPC, multi-parametric model predictive control): (i) the reformulation of the linear MPC problem with quadratic performance index into a quadratic programming problem, free of equality constraints, via successive substitution of state and output variables, over the finite horizon and (ii) the explicit/multi-parametric solution of the quadratic programming problem as a function of parameters. In (i) we show the steps from a transfer function into an MPC formulation. The latter is shown to be a multi-parametric quadratic programming (mp-QP) problem where the optimization variables are the control inputs and the uncertainty stems from the current state of the process/system at hand, for which we assume full measurement, are available. In (ii) we use the algorithms presented in Chapter 3 to derive the explicit control solution as a function of the initial states. A typical and widely addressed linear-quadratic regulator example is used and extensions to set-point tracking and disturbance rejection are briefly presented. Last, we extend this practice into the general case of receding horizon optimization problems with linear dynamics.

10.2 From Transfer Functions to Discrete Time State-Space Models

Consider the following second order transfer function correlating the systems input $u(s)$ and output $y(s)$ in the frequency domain:

$$G(s) = \frac{y(s)}{u(s)} = \frac{k}{s^2 + as + b}. \tag{10.1}$$

The system of Eq. (10.1) can be rewritten as follows:

$$G(s) = N(s)D(s), \tag{10.2a}$$

$$N(s) = \frac{y(s)}{\chi(s)} = k, \tag{10.2b}$$

$$D(s) = \frac{\chi(s)}{u(s)} = \frac{1}{s^2 + as + b}. \tag{10.2c}$$

By rewriting the term $D(s)$ in differential form in the time domain through an inverse Laplace transform, we get:

$$\ddot{\chi}(t) + a\dot{\chi}(t) + b\chi(t) = u(t). \tag{10.3}$$

Consequently, by defining $x(t) = [\dot{\chi}(t)\chi(t)]^T$, we express the second order transfer function of Eq. (10.1) in the continuous time state-space model of Eq. (10.4):

$$\begin{cases} \dot{x}(t) = \begin{bmatrix} -a & -b \\ 1 & 0 \end{bmatrix} x(t) + \begin{bmatrix} 1 \\ 0 \end{bmatrix} u(t), \\ y(t) = \begin{bmatrix} 0 & k \end{bmatrix} x(t). \end{cases} \tag{10.4}$$

In general terms, we define the continuous time state-space model of the form:

$$\begin{cases} \dot{x}(t) = \hat{A}x(t) + \hat{B}u(t), \\ y(t) = Cx(t). \end{cases} \tag{10.5}$$

There are a few things worth noting regarding the transformation of the transfer function to a continuous time state-space model. There are a total of four *classic* transformations called "canonical forms" of the state-space model, namely, the observer, controller, observability, and controllability forms, which are useful in unique ways. The freedom in choosing and defining the state vector $x(t)$ is what determines which form will be generated. Previously, we generated the controller form since the focus of the chapter is the derivation of mp-MPCs. The interested reader is referred to [2] for a complete analysis of the transformations and forms of transfer functions and state-space models. The approach presented here can be generalized to a transfer function of order n and is not restricted to order 2.

Depending on the numerator of the transfer function $N(s)$, there may be feedthrough of the input to the output, i.e. $y(t) = Cx(t) +$

$Du(t)$. An example of this can be seen in a transfer function of the form:

$$G(s) = \frac{y(s)}{u(s)} = \frac{b_0 s^3 + b_1 s^2 + b_2 s + b_3}{s^3 + a_1 s^2 + a_2 s + a_3}. \tag{10.6}$$

By defining $D = b_0$ and $\beta_i = b_i - b_0 a_i$ we get:

$$G(s) = G_1(s) + D, \tag{10.7a}$$

$$G_1(s) = \frac{y_1(s)}{u(s)} = \frac{\beta_1 s^2 + \beta_2 s + \beta_3}{s^3 + a_1 s^2 + a_2 s + a_3} = N_1(s) D_1(s), \tag{10.7b}$$

$$N_1(s) = \frac{y_1(s)}{\chi(s)} = \beta_1 s^2 + \beta_2 s + \beta_3, \tag{10.7c}$$

$$D_1(s) = \frac{\chi(s)}{u(s)} = \frac{1}{s^2 + as + b}. \tag{10.7d}$$

By defining $x(t) = [\ddot{\chi}(t)\dot{\chi}(t)\chi(t)]^T$, we can show that the corresponding control canonical form of $G_1(s)$ is defined, like previously, as

$$\begin{cases} \dot{x}(t) = \begin{bmatrix} -a_1 & -a_2 & -a_3 \\ 1 & 0 & 0 \\ 0 & 1 & 0 \end{bmatrix} x(t) + \begin{bmatrix} 1 \\ 0 \\ 0 \end{bmatrix} u(t), \\ y_1(t) = \begin{bmatrix} \beta_1 & \beta_2 & \beta_3 \end{bmatrix} x(t). \end{cases} \tag{10.8}$$

Observing from Eq. (10.7a) that $G(s) = G_1(s) + Du(s)$, we derive the continuous time state-space model with feedthrough as

$$\begin{cases} \dot{x}(t) = \begin{bmatrix} -a_1 & -a_2 & -a_3 \\ 1 & 0 & 0 \\ 0 & 1 & 0 \end{bmatrix} x(t) + \begin{bmatrix} 1 \\ 0 \\ 0 \end{bmatrix} u(t), \\ y(t) = \begin{bmatrix} b_1 - b_0 a_1 & b_2 - b_0 a_2 & b_3 - b_0 a_3 \end{bmatrix} x(t) + b_0 u(t). \end{cases} \tag{10.9}$$

Hereon the representation $\dot{x}(t) = \hat{A}x(t) + \hat{B}u(t)$, $y(t) = Cx(t) + Du(t)$, with or without feedthrough, will be used to denote the continuous time, linear, time invariant, state-space model representation, and its corresponding matrices.

In order to discretize the continuous state-space model, we need to define matrices $A(T)$ and $B(T)$ such that

$$\begin{cases} x((k+1)T) = A(T)x(kT) + B(T)u(kT), \\ y(kT) = Cx(kT) + Du(kT), \end{cases} \tag{10.10}$$

where T is the discretization time. The question then becomes how to obtain $A(T)$ and $B(T)$ in Eq. (10.10). The aforementioned matrices are a function of the matrices $\hat{A}$ and $\hat{B}$ and the discretization time T. The latter is a choice based on numerous factors. In this case, a uniform discretization time step is employed. Thus, we write

$$\dot{x}(t) = \hat{A}x(t) + \hat{B}u(t), \tag{10.11a}$$

$$-\hat{A}x(t) + \dot{x}(t) = \hat{B}u(t). \tag{10.11b}$$

We observe that $\frac{d(e^{-\hat{A}t}x(t))}{dt} = -\hat{A}\ e^{-\hat{A}t}x(t) + e^{-\hat{A}t}\dot{x}$. We also observe that $e^{-\hat{A}t}$ and $-\hat{A}$ are mirror matrices, i.e. $e^{-\hat{A}t}(-\hat{A}) = -\hat{A}\ e^{-\hat{A}t}$ for every $\hat{A}$ and every t. Therefore, after multiplying both sides of Eq. (10.11b) with $e^{-\hat{A}t}$, we write

$$-\hat{A}\ e^{-\hat{A}t}x(t) + e^{-\hat{A}t}\ \dot{x}(t) = e^{-\hat{A}t}\ \hat{B}u(t), \tag{10.12a}$$

$$\frac{d(e^{-\hat{A}t}\ x(t))}{dt} = e^{-\hat{A}t}\ \hat{B}u(t). \tag{10.12b}$$

By integrating from the kth time step to the $(k+1)$th time step and assuming the input $u(t)$ is fixed during the time step, i.e. $u(t = kT) = u(t = (k+1)T)$, we see that

$$\int_{kT}^{(k+1)T} \frac{d(e^{-\hat{A}t}x(t))}{dt} dt = \int_{kT}^{(k+1)T} e^{-\hat{A}t}\ \hat{B}u(t)dt, \tag{10.13a}$$

$$\int_{kT}^{(k+1)T} \frac{d(e^{-\hat{A}t}x(t))}{dt} dt = \int_{kT}^{(k+1)T} e^{-\hat{A}t}\ dt\hat{B}u(kT), \tag{10.13b}$$

$$\begin{aligned} &e^{-\hat{A}(k+1)T}\ x((k+1)T) - e^{-\hat{A}kT}\ x(kT) \\ &= -\hat{A}^{-1}(e^{-\hat{A}(k+1)T} - e^{-\hat{A}kT})\hat{B}u(kT). \end{aligned} \tag{10.13c}$$

It follows that by rearranging the terms and multiplying both sides of the equation with $e^{-\hat{A}(k+1)T}$, we get

$$x((k+1)T) = \overbrace{e^{\hat{A}T}}^{A(T)} x(kT) + \overbrace{\hat{A}^{-1}(e^{\hat{A}T} - I)\hat{B}}^{B(T)} u(kT). \quad (10.14)$$

Note that the algebraic equation $y(t) = Cx(t) + Du(t)$ is not affected by the discretization and is simply transformed into $y(kT) = Cx(kT) + Du(kT)$. Also, for simplicity, by defining matrices $A = A(T)$, $B = B(T)$, and vectors $v(kT) = v_k$ for all vectors x, u, and y the final form of the discrete time, linear, time invariant state-space model reads

$$\begin{cases} x_{k+1} = Ax_k + Bu_k, \\ y_k = Cx_k + Du_k. \end{cases} \quad (10.15)$$

The model of Eq. (10.15) will be used hereon to demonstrate how the MPC problem results into a multi-parametric programming formulation.

10.3 From Discrete Time State-Space Models to Multi-parametric Programming

Consider the problem of regulating to the origin the discrete-time, linear, time invariant system of Eq. (10.15): under the following linear constraints:

$$x_{\min} \le x_k \le x_{\max}, \quad y_{\min} \le y_k \le y_{\max}, \quad u_{\min} \le u_k \le u_{\max}. \quad (10.16)$$

The MPC problem corresponding to regulating a discrete-time, linear, time invariant system to its origin is called a linear quadratic regulator (LQR) and is presented in Eq. (10.17) for the case of a quadratic performance index.[2]

2 We omit the term Du_k in the output for simplicity but the feedthrough case follows similarly.

$$
\begin{aligned}
&\underset{u}{\text{Minimize}} \; x_N^T P x_N + \sum_{k=0}^{N-1} x_k^T Q x_k + \sum_{k=0}^{M-1} u_k^T R u_k \\
&\text{Subject to} \begin{cases} x_{k+1} = Ax_k + Bu_k \\ y_k = Cx_k \end{cases} && \forall k \in [0, M-1] \\
&\qquad\qquad\;\; \begin{cases} x_{k+1} = Ax_k + Bu_{M-1} \\ y_k = Cx_k \end{cases} && \forall k \in [M, N-1] \\
&\qquad\qquad\;\; x_{\min} \le x_k \le x_{\max}, && \forall k \in [0, N] \\
&\qquad\qquad\;\; y_{\min} \le y_k \le y_{\max}, && \forall k \in [0, N] \\
&\qquad\qquad\;\; u_{\min} \le u_k \le u_{\max}, && \forall k \in [0, M-1] \\
&\qquad\qquad\;\; A_T x_N \le b_T,
\end{aligned}
\tag{10.17}
$$

where N is the prediction horizon, $M \le N$ the control horizon, $Q \succeq 0$ the weight for the states, $R \succeq 0$ the weight for the inputs, P the final state weight calculated as discussed in [1], $x_{\min}$ and $x_{\max}$ the lower and upper bound on x_k, respectively, $u_{\min}$ and $u_{\max}$ the lower and upper bound on u_k, respectively, and $y_{\min}$ and $y_{\max}$ the lower and upper bound on y_k, respectively. Note that the problem formulation includes a set of linear constraints corresponding to the final state values x_N, known as the terminal set.

Remark 10.1 It is equally possible to use a linear performance index such as the $1/\infty$-norm.

It is possible and often encountered to consider an equal control and prediction horizon, i.e. M = N in eq. (10.17). In this case the LQR problem is shown in Eq. (10.18).

$$
\begin{aligned}
&\underset{u}{\text{Minimize}} \; x_N^T P x_N + \sum_{k=0}^{N-1} x_k^T Q x_k + u_k^T R u_k \\
&\text{Subject to} \begin{cases} x_{k+1} = Ax_k + Bu_k \\ y_k = Cx_k \end{cases} && \forall k \in [0, N-1] \\
&\qquad\qquad\;\; x_{\min} \le x_k \le x_{\max}, && \forall k \in [0, N] \\
&\qquad\qquad\;\; y_{\min} \le y_k \le y_{\max}, && \forall k \in [0, N] \\
&\qquad\qquad\;\; u_{\min} \le u_k \le u_{\max}, && \forall k \in [0, N-1] \\
&\qquad\qquad\;\; A_T x_N \le b_T.
\end{aligned}
\tag{10.18}
$$

Based on Eqs. (10.15) and (10.17) every state vector for which $k \geq 1$ can be reformulated as follows:

$$\begin{aligned} x_k = A^k x_0 + & \sum_{k \in [1,M],\ i=0}^{k-1} A^{k-i-1} B u_i + \\ + & \sum_{k \in [M+1,N],\ j=M}^{k} A^{N-j-1} B u_{M-1}, \forall k \in [1,N]. \end{aligned} \tag{10.19}$$

Equivalently, for $N = M$, based on Eqs. (10.15) and (10.18) every state vector for which $k \geq 1$ can be reformulated as follows:

$$x_k = A^k x_0 + \sum_{i=0}^{k-1} A^{k-i-1} B u_i, \quad \forall k \in [1,N]. \tag{10.20}$$

Equations (10.19) and (10.20) are linear in x_0 and u_i, $\forall i \in [0, M-1]$. A simple example of its application is given in Table 10.1 for $N = 4, M = 2$ and $N = 4, M = 4$.

The states of the system can therefore be expressed as a linear function of the form of Eq. (10.21). From this point hereon, we will only consider the case where the prediction and control horizons are equal.

$$\begin{aligned} x_k = &\ A_r^k x_0 + B_r^k u, \\ u = &\ [u_1^T, u_2^T, \dots, u_{k-1}^T]^T, \end{aligned} \tag{10.21}$$

Table 10.1 State expressions as a function of the initial state values and inputs.

Horizons	x_k as a linear function of x_0 and u	
$N = 4, M = 2$	$k = 1$	$x_1 = Ax_0 + Bu_0$
	$k = 2$	$x_2 = A^2 x_0 + ABu_0 + Bu_1$
	$k = 3$	$x_3 = A^3 x_0 + A^2 Bu_0 + (AB + B)u_1$
	$k = 4$	$x_4 = A^4 x_0 + A^3 Bu_0 + (A^2 B + AB + B)u_1$
$N = M = 4$	$k = 1$	$x_1 = Ax_0 + Bu_0$
	$k = 2$	$x_2 = A^2 x_0 + ABu_0 + Bu_1$
	$k = 3$	$x_3 = A^3 x_0 + A^2 Bu_0 + ABu_1 + Bu_2$
	$k = 4$	$x_4 = A^4 x_0 + A^3 Bu_0 + A^2 Bu_1 + ABu_2 + Bu_3$

where A_r^k and B_r^k are the A_r and B_r fixed, concatenated matrices corresponding to the linear reformulation based on the original discrete-time, linear, time invariant system as shown in Eqs. (10.19) and (10.20) for the kth state. Note that the states of the system are a linear function of the initial state value vector x_0 and the control variables vector u. Also note that the matrix B_r^k is matrix with as many lines as the numbers of states and as many columns as the number of input variables multiplied by the control horizon. Therefore, it follows that the B_r^k matrix is a sparse matrix with all zero elements from the kth column to the $N-1$th column.

Equivalently, the outputs of the system can be expressed as linear functions by substitution from Eq. (10.17) shown in Eq. (10.22):

$$\begin{aligned} y_k &= CA_{r,k}x_0 + CB_{r,k}u, \\ u &= [u_1^T, u_2^T, \dots, u_{k-1}^T]^T. \end{aligned} \tag{10.22}$$

By substituting Eqs. (10.21) and (10.22) into the original MPC formulation of Eq. (10.18), we get only linear inequality constraints based on the upper and lower bounds. The original quadratic objective function remains quadratic and three quadratic terms corresponding to (i) the quadratic term of the control variables, (ii) the bilinear term between the control variables and the initial states, and (iii) the quadratic term of the initial states are generated.[3] In order for this to be more comprehensive, we consider a simple example based on Eqs. (10.18), (10.19), and (10.22).

After rearranging the terms, the resulting quadratic programming problem is presented in Eq. (10.23):

$$\underset{u}{\text{Minimize}}\; u^T \underbrace{\left[B_r^{N^T} P B_r^N + \sum_{i=0}^{N-1} (B_r^{i^T} Q B_r^i) + R_d \right]}_{H} u$$

3 Similarly for a linear performance index, the objective function remains linear.

$$
+u^T \underbrace{\left[B_r^{N^T}(P+P^T)A_r^N + \sum_{i=0}^{N-1}(B_r^{i^T}(Q+Q^T)A_r^i) \right]}_{Z} x_0
$$

$$
+x_0^T \underbrace{\left[A_r^{N^T} P A_r^N + \sum_{i=0}^{N-1}(A_r^{i^T} Q A_r^i) \right]}_{\hat{M}} x_0
$$

$$
\text{Subject to} \quad \underbrace{\begin{bmatrix} B_r^i \\ -B_r^i \\ CB_r^i \\ -CB_r^i \\ \mathbf{I} \\ -\mathbf{I} \\ A_T B_r^N \end{bmatrix}}_{G} u \le \underbrace{\begin{bmatrix} x_{max} \\ -x_{min} \\ y_{max} \\ -y_{min} \\ u_{max} \\ -u_{min} \\ b_T \end{bmatrix}}_{W} + \underbrace{\begin{bmatrix} -A_r^i \\ A_r^i \\ -CA_r^i \\ CA_r^i \\ \mathbf{0} \\ \mathbf{0} \\ -A_T A_r^N \end{bmatrix}}_{S} x_0
$$

$$
\underbrace{\begin{bmatrix} \mathbf{I} \\ -\mathbf{I} \\ C \\ -C \end{bmatrix}}_{\mathrm{CR}_A} x_0 \le \underbrace{\begin{bmatrix} x_{max} \\ -x_{min} \\ y_{max} \\ -y_{min} \end{bmatrix}}_{\mathrm{CR}_b}, \tag{10.23}
$$

where H is the quadratic term for the control variables, Z the bilinear term between the control variables and initial states, $\hat{M}$ the quadratic term for the initial states, $\mathbf{I}$ the identity matrix, and $\mathbf{0}$ a zero matrix, all of appropriate dimensions. Note that the matrix R_d is a matrix that repeats the R matrix on its diagonal N times.

Remark 10.2 It can be proven that for any matrices A and B of the linear, discrete time, time invariant state-space model of Eq.10.15 the resulting quadratic problem of Eq. (10.23) is convex in u (i.e. $z^T H z \ge 0,\ \forall z \in \mathbf{R}^n$) as long as the matrices Q and R are convex and symmetric.

Remark 10.3 The term $\hat{M}$ does not affect the outcome of the optimization problem; therefore it can be omitted. The term $x_0^T \hat{M} x_0$

corresponds to a fixed value as soon as the initial state values realize; therefore the value of the optimal u is not affected by it.

In the cases where the performance index of the MPC problem (Eq. (10.17)) is linear via a $1/\infty$-norm formulation, it can be equivalently shown that

Remark 10.4 The objective function of the optimization problem remains linear. The two terms that result from the reformulation correspond to (i) the linear term of the control variables and (ii) the linear term of the initial states.

Remark 10.5 The reformulation of the constraints of the problem remains the same. A set of constraints may be added to preserve the properties of the infinite norm.

Remark 10.6 The corresponding optimization problem becomes a linear programming problem under linear constraints.

$$\begin{aligned} &\underset{u}{\text{Minimize}} \ u^T H u + u^T Z x_0 \\ &\text{Subject to} \ Gu \leq W + Sx_0 \\ &\qquad\qquad \mathrm{CR}_A x_0 \leq \mathrm{CR}_b. \end{aligned} \tag{10.24}$$

Equation (10.24) corresponds to a convex mp-QP problem where the parameters are the initial state values x_0 and the control variables u are the decision variables. Equation (10.24) is exactly equivalent to Eq. (3.2) and can be solved with the methods presented in Chapter 4. The next section of this chapter tackles an example of an LQR problem.

10.4 Explicit LQR – An Example of mp-MPC

10.4.1 Problem Formulation and Solution

Consider the discrete-time, linear, time invariant model of Eq. (10.15).[4] In this example, we consider the following system

4 In order to (i) avoid the complexity associated with the transformations and discretization and (ii) to underline that not all state-space models stem from transfer function we begin the example from the point of the discrete-time, linear time invariant model.

dynamics (Eq. (10.25)), box constraints (Eq. (10.26)), objective function weights (Eq. (10.27)), horizons (Eq. (10.28)), and terminal constraints (Eq. (10.29)):

$$A = \begin{bmatrix} 1 & 1 \\ 0 & 1 \end{bmatrix}, \; B = \begin{bmatrix} 0 \\ 1 \end{bmatrix}, \; C = \begin{bmatrix} 1 & 2 \end{bmatrix}, \; D = \begin{bmatrix} 0 \end{bmatrix}. \tag{10.25}$$

$$\begin{aligned} &\text{Bounds on states:} && \begin{bmatrix} -10 \\ -10 \end{bmatrix} \leq x_k \leq \begin{bmatrix} 10 \\ 10 \end{bmatrix}, && \forall k \in [0, N], \\ &\text{Bounds on inputs:} && -1 \leq u_k \leq 1, && \forall k \in [0, N-1], \\ &\text{Bounds on outputs:} && -25 \leq y_k \leq 25, && \forall k \in [0, N], \end{aligned} \tag{10.26}$$

$$Q = \begin{bmatrix} 1 & 0 \\ 0 & 1 \end{bmatrix}, \; R = 0.01, \tag{10.27}$$

$$\begin{aligned} &\text{Prediction horizon:} && N = 10, \\ &\text{Control horizon:} && M = 10. \end{aligned} \tag{10.28}$$

$$A_T = \begin{bmatrix} 0.6136 & 1.6099 \\ -0.3742 & -0.3682 \\ -0.6136 & -1.6099 \\ 0.3742 & 0.3682 \end{bmatrix}, \; b_T = \begin{bmatrix} 1 \\ 1 \\ 1 \\ 1 \end{bmatrix}. \tag{10.29}$$

Given Eqs. (10.25)–(10.29) and the MPC formulation in Eq. (10.18), the MPC problem is presented in Eq. (10.30).

$$\begin{aligned} \underset{u}{\text{Minimize}} \; & x_{10}^T \begin{bmatrix} 2.6235 & 1.6296 \\ 1.6296 & 2.6457 \end{bmatrix} x_{10} + \sum_{k=0}^{9} x_k^T \begin{bmatrix} 1 & 0 \\ 0 & 1 \end{bmatrix} x_k + u_k^T 0.01 u_k \\ \text{Subject to} \; & \begin{cases} x_{k+1} = \begin{bmatrix} 1 & 1 \\ 0 & 1 \end{bmatrix} x_k + \begin{bmatrix} 0 \\ 1 \end{bmatrix} u_k \\ y_k = \begin{bmatrix} 1 & 2 \end{bmatrix} x_k \end{cases}, \quad \forall k \in [0, 9] \\ & \begin{bmatrix} -10 \\ -10 \end{bmatrix} \leq x_k \leq \begin{bmatrix} 10 \\ 10 \end{bmatrix}, \quad \forall k \in [0, 10] \\ & -25 \leq y_k \leq 25, \quad \forall k \in [0, 10] \\ & -1 \leq u_k \leq 1, \quad \forall k \in [0, 9] \\ & \begin{bmatrix} 0.6136 & 1.6099 \\ -0.3742 & -0.3682 \\ -0.6136 & -1.6099 \\ 0.3742 & 0.3682 \end{bmatrix} x_{10} \leq \begin{bmatrix} 1 \\ 1 \\ 1 \\ 1 \end{bmatrix}. \end{aligned} \tag{10.30}$$

Following the procedure described by Eqs. (10.19)–(10.22), we derive the mp-QP formulation corresponding to the mp-MPC in the form of Eq. (10.24). Equations (A.2)–(A.6) in Appendix A, define the corresponding mp-QP matrices. Note that this is the minimal problem representation, without the quadratic term of the parameters on the objective function and after the removal of redundant constraints (if any).[5]

Having brought the problem into the standard mp-QP format and by applying the procedures and algorithms in Chapters 3 and 4, we can acquire the explicit solution for the constrained LQR problem.

10.4.2 Results and Validation

The solution to the mp-QP of Eqs. (10.24) and (A.2)–(A.6) yields a solution of 115 critical regions using the connected graph algorithm (Figure 10.2). Every critical region has a distinct parametric optimal function set for the 10 optimization variables u. In Table 10.2 we present the full set of critical regions graphically and the analytical expressions for 2 critical regions. Note that the definition of the critical region where the input is optimal is denoted by the constraints of the form $\mathrm{CR} = \mathrm{CR}_A(i)x_0 \leq \mathrm{CR}\ (i)_b$, the regions active set is given by k the corresponding optimal input is given in the form of $u(x_0) = A(i)x_0 + b(i)$, the Lagrange multipliers (also referred to as co-states in the context of MPC) are given in the form of $\lambda(x_0) = A_{\lambda,i}x_0 + b_{\lambda,i}$, and the objective function value is given in the form $z(x_0) = x_0^T Q(i)x_0 + c_i^T x_0 + d_i$, where i is the critical region number.

There are a few things worth noting regarding the solution of the mp-MPC problem. The unconstrained region, corresponding in this case in CR_1, corresponds to the unconstrained MPC solution and satisfies the equation $u_k = -Kx_k$, $\forall k \in 0, 9$. The optimal control feedback law for this problem can be calculated to $K = \begin{bmatrix}0.61 & 1.61\end{bmatrix}$based on our choice of Q and R matrices and the system dynamics. The solution of the unconstrained region optimal input action is shown in Table 10.3 along with the

5 Usually, mp-MPC problems suffer from a large number of redundant constraints. The removal of the constraints a priori can help with the solution procedure significantly.

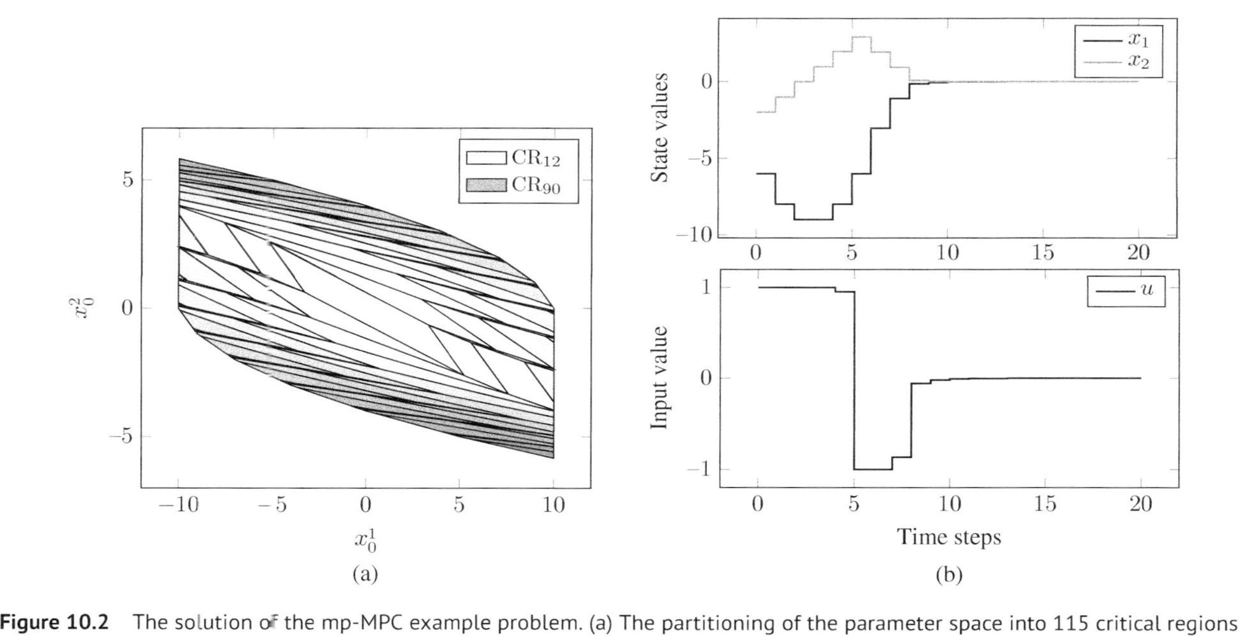

Figure 10.2 The solution of the mp-MPC example problem. (a) The partitioning of the parameter space into 115 critical regions is shown, each of which is associated with a different set of active constraints k_i. The legend shows only 2 out of the 115 critical regions to simplify the figure. The full solutions to the same regions is presented in Table 10.2. (b) The linear, time invariant, state-space, model is simulated in a closed loop fashion. In this particular snapshot, the initial state values have been chosen to be $x_0^1 = -6$ and $x_0^2 = -2$. Observe how the controller is able to stabilize the system around the origin and achieve a final value for the state $x_f^1 = 0$ and $x_f^2 = 0$.

Table 10.2 The partial analytical solution of the mp-MPC problem.

$$\mathrm{CR}_{12}\left\{\begin{array}{l}
\mathrm{CR} = \begin{bmatrix} 0.250.97 \\ 0.710.71 \\ -0.25-0.97 \\ -0.71-0.7 \\ 10 \end{bmatrix} x_0 \le \begin{bmatrix} 0.09 \\ 4.51 \\ 1.34 \\ -3.03 \\ 10 \end{bmatrix} \\
k = \{24, 25\} \\
u(x_0) = \begin{bmatrix} -0.35-1.35 \\ 0.00E+000.00E+00 \\ 0.00E+000.00E+00 \\ 0.370.37 \\ -0.01-0.01 \\ -0.01-0.01 \\ 00 \\ 00 \\ 00 \\ 00 \end{bmatrix} x_0 + \begin{bmatrix} -0.87 \\ 1 \\ 1 \\ -1.37 \\ 0.14 \\ 0.06 \\ 0.02 \\ 0.01 \\ 0 \\ 0 \end{bmatrix} \\
\lambda_k(x_0) = \begin{bmatrix} 0.840.84 \\ 1.071.06 \end{bmatrix} x_0 + \begin{bmatrix} -3.61 \\ -3.69 \end{bmatrix} \\
z(x_0) = x_0^T \begin{bmatrix} 1.891.89 \\ 1.891.9 \end{bmatrix} x_0 + \begin{bmatrix} -1.92 \\ -1.9 \end{bmatrix}^T x_0 + 3.65
\end{array}\right.$$

$$\mathrm{CR}_{90}\left\{\begin{array}{l}
\mathrm{CR} = \begin{bmatrix} -0.24 & -0.97 \\ -0.14 & -0.99 \\ -0.1 & -0.99 \\ 0.14 & 0.99 \end{bmatrix} x_0 \le \begin{bmatrix} 3.88 \\ 3.87 \\ 3.89 \\ -3.57 \end{bmatrix} \\
k = \{14, 15, 21, 22, 23, 24, 25, 27\} \\
u(x_0) = \begin{bmatrix} 0.00E+00 & 0.00E+00 \\ 0 & 0 \\ 0 & 0 \\ 0 & 0 \\ 0 & 0 \\ 0 & 0 \\ -0.35 & -3.43 \\ 0 & 0 \\ 0 & 0 \\ 0.37 & 2.6 \end{bmatrix} x_0 + \begin{bmatrix} 1 \\ 1 \\ 1 \\ 1 \\ 1 \\ 1 \\ -12.43 \\ -1 \\ -1 \\ 9.18 \end{bmatrix} \\
\lambda_k(x_0) = \begin{bmatrix} -1.07 & -7.49 \\ -0.84 & -5.91 \\ -3.78 & -28.49 \\ -9.55 & -68.91 \\ -17.33 & -119.33 \\ -27.1 & -177.74 \\ -38.87 & -242.16 \\ -52.65 & -310.58 \end{bmatrix} x_0 + \begin{bmatrix} -26.11 \\ -21.31 \\ -93.46 \\ -214.66 \\ -353.85 \\ -503.04 \\ -656.24 \\ -809.43 \end{bmatrix} \\
z(x_0) = x_0^T \begin{bmatrix} 7.89 & 34.21 \\ 34.21 & 189.5 \end{bmatrix} x_0 + \begin{bmatrix} 151.19 \\ 960.6 \end{bmatrix}^T x_0 + 1339.05
\end{array}\right.$$

The term CR corresponds to the critical region definition, the term k to the active set, the term $u(x_0)$ to the optimal parametric input function, the term $\lambda(x_0)$ to the Lagrange multipliers of the active constraints, and the term $z(x_0)$ to the parametric objective function.

Table 10.3 Comparison of optimal unconstrained feedback law (via $x_k = (A - BK)^k x_0, \ \forall k \in [0, 9]$) and optimal input in the unconstrained mp-MPC critical region CR_1.

Horizon step (k)	u_k as a linear function of x_0	
	Optimal unconstrained feedback law	Parametric solution for CR_1
$k = 0$	$u_0 = \begin{bmatrix} -0.61 & -1.61 \end{bmatrix} x_0$	$u_0 = \begin{bmatrix} -0.61 & -1.61 \end{bmatrix} x_0$
$k = 1$	$u_1 = -Kx_1 = \begin{bmatrix} 0.37 & 0.37 \end{bmatrix} x_0$	$u_1 = \begin{bmatrix} 0.37 & 0.37 \end{bmatrix} x_0$
$k = 2$	$u_2 = -Kx_2 = \begin{bmatrix} 0.15 & 0.15 \end{bmatrix} x_0$	$u_2 = \begin{bmatrix} 0.15 & 0.15 \end{bmatrix} x_0$
$k = 3$	$u_3 = -Kx_3 = \begin{bmatrix} 0.06 & 0.06 \end{bmatrix} x_0$	$u_3 = \begin{bmatrix} 0.06 & 0.06 \end{bmatrix} x_0$
$k = 4$	$u_4 = -Kx_4 = \begin{bmatrix} 0.02 & 0.02 \end{bmatrix} x_0$	$u_4 = \begin{bmatrix} 0.02 & 0.02 \end{bmatrix} x_0$
$k = 5$	$u_5 = -Kx_5 = \begin{bmatrix} 0.01 & 0.01 \end{bmatrix} x_0$	$u_5 = \begin{bmatrix} 0.01 & 0.01 \end{bmatrix} x_0$
$k = 6$	$u_6 = -Kx_6 = \begin{bmatrix} 0 & 0 \end{bmatrix} x_0$	$u_6 = \begin{bmatrix} 0 & 0 \end{bmatrix} x_0$
$k = 7$	$u_7 = -Kx_7 = \begin{bmatrix} 0 & 0 \end{bmatrix} x_0$	$u_7 = \begin{bmatrix} 0 & 0 \end{bmatrix} x_0$
$k = 8$	$u_8 = -Kx_8 = \begin{bmatrix} 0 & 0 \end{bmatrix} x_0$	$u_8 = \begin{bmatrix} 0 & 0 \end{bmatrix} x_0$
$k = 9$	$u_9 = -Kx_9 = \begin{bmatrix} 0 & 0 \end{bmatrix} x_0$	$u_9 = \begin{bmatrix} 0 & 0 \end{bmatrix} x_0$

correlation to the optimal feedback law K. The optimal input outside the unconstrained region is such that it guarantees that no constraints are violated.

10.5 Size of the Solution and Online Computational Effort

The two major advantages of multi-parametric/explicit MPC are its ability to provide a "map of solutions" a priori and its effortless online applicability. In the first case, the fact that the entire control problem is solved offline provides great insight regarding the effect that the initial state vector values have on the optimal control action. This helps the control developer understand the control behavior and guarantee the optimality of the solution. Furthermore, in the presence of (measured) disturbances within the control scheme, the same can be guaranteed.

The effortless online applicability is a direct result of the nature of the MPC problem. The linear or quadratic formulation of the problem guarantees that the optimal control action is linear with respect to the parameters, in this case, the initial state vector values. Furthermore, the critical regions for which an optimal action remains optimal are also a polytope. The fact that the explicit expression of the optimal control action is linear results in fast MPC computations without the need of solving an optimization problem online. It is fair to say that the major burden of the optimization based MPC application has therefore been alleviated.[6]

The computationally most expensive step with respect to the application of multi-parametric/explicit MPC is the identification of the critical region within which the parameter vector lies. In essence, it is a set membership test of a series of disjoint polytopes whose union might be convex or non-convex, depending on whether the multi-parametric programming problem features binary variables. While it is possible to exhaustively enumerate all

6 Note that this claim refers mainly to alleviating the necessity of optimization hardware equipment for the application of optimization based MPC as the explicit solution enables the use of MPC-on-a-chip as described in [5].

possible polytopes, such an approach becomes computationally problematic in the case of larger systems with potentially thousands of critical regions. In addition, from a practical view point, it is equally important to provide an upper bound on the worst-case scenario for the point location problem.[7] Thus, starting with works in [6, 7], several researchers have considered the design of efficient algorithms [8–25].

References

1 Bemporad, A., Morari, M., Dua, V., and Pistikopoulos, E.N. (2002) The explicit linear quadratic regulator for constrained systems. *Automatica*, 38 (1), 3–20, doi: 10.1016/S0005-1098(01)00174-1. URL http://www.sciencedirect.com/science/article/pii/S0005109801001741.

2 Rawlings, J.B., Mayne, D.Q., and M., D.M. (2017) *Model predictive control: theory and design*, Nob Hill Publishing, Madison, WI, 2nd edn, 978-0975937730.

3 Kouvaritakis, B. and Cannon, M. (2016) *Model predictive control: classical, robust and stochastic*, *Advanced textbooks in control and signal processing*, Springer, Switzerland.

4 Raković, S.V. and Levine, W.S. (2019) *Handbook of model predictive control*, Springer International Publishing, Cham, 1st edn. URL https://doi.org/10.1007/978-3-319-77489-3.

5 Pistikopoulos, E.N., Dominguez, L., Panos, C., Kouramas, K., and Chinchuluun, A. (2012) Theoretical and algorithmic advances in multi-parametric programming and control. *Computational Management Science*, 9 (2), 183–203, doi: 10.1007/s10287-012-0144-4. URL http://dx.doi.org/10.1007/s10287-012-0144-4.

6 Johansen, T.A. and Grancharova, A. (2003) Approximate explicit constrained linear model predictive control via orthogonal search tree. *IEEE Transactions on Automatic Control*, 48 (5), 810–815, doi: 10.1109/TAC.2003.811259.

7 If the sampling time of a system is 1 μs, but the point location of the explicit MPC controller may require up to 5 μs, the explicit MPC controller cannot be applied in practice.

7 Tøndel, P., Johansen, T.A., and Bemporad, A. (2003) Evaluation of piecewise affine control via binary search tree. *Automatica*, 39 (5), 945–950, doi: 10.1016/S0005-1098(02)00308-4. URL http://www.sciencedirect.com/science/article/pii/S0005109802003084.

8 Borrelli, F., Baotic, M., Bemporad, A., and Morari, M. (2001) Efficient on-line computation of constrained optimal control, in *Proceedings of the 40th IEEE Conference on Decision and Control, 2001*, vol. 2, pp. 1187–1192, doi: 10.1109/.2001.981046.

9 Jones, C.N., Grieder, P., and Raković, S.V. (2006) A logarithmic-time solution to the point location problem for parametric linear programming. *Automatica*, 42 (12), 2215–2218, doi: 10.1016/j.automatica.2006.07.010. URL http://www.sciencedirect.com/science/article/pii/S0005109806003074.

10 Spjotvold, J., Rakovic, S.V., Tondel, P., and Johansen, T.A. (2006) Utilizing reachability analysis in point location problems, in *2006 45th IEEE Conference on Decision and Control*, pp. 4568–4569, doi: 10.1109/CDC.2006.377580.

11 Christophersen, F.J., Kvasnica, M., Jones, C.N., and Morari, M. (2007) Efficient evaluation of piecewise control laws defined over a large number of polyhedra, in *Proceedings of the European Control Conference*, Kos, Greece.

12 Wang, Y., Jones, C., and Maciejowski, J. (2007) Efficient point location via subdivision walking with application to explicit MPC, in *Proceedings of the European Control Conference*, Kos, Greece (2–5 July 2007).

13 Baotic, M., Borrelli, F., Bemporad, A., and Morari, M. (2008) Efficient on-line computation of constrained optimal control. *SIAM Journal on Control and Optimization*, 47 (5), 2470–2489. URL http://epubs.siam.org/doi/pdf/10.1137/060659314.

14 Grancharova, A. and Johansen, T.A. (2009) Approaches to explicit nonlinear model predictive control with reduced partition complexity, in *2009 European Control Conference (ECC)*, Budapest, Hungary (23–26 August 2009), pp. 2414–2419.

15 Fuchs, A., Axehill, D., and Morari, M. (2010) On the choice of the linear decision functions for point location in polytopic data sets - application to explicit MPC, in *IEEE Conference on Decision and Control*, Atlanta, USA, pp. 5283–5288. URL http://

control.ee.ethz.ch/index.cgi?page=publications;action=details;id=3656.

16 Monnigmann, M. and Kastsian, M. (2011) Fast explicit model predictive control with multiway trees, in *World Congress*, IFAC, Elsevier, IFAC proceedings volumes, pp. 1356–1361, doi: 10.3182/20110828-6-IT-1002.00686.

17 Bayat, F., Johansen, T.A., and Jalali, A.A. (2011) Using hash tables to manage the time-storage complexity in a point location problem: application to explicit model predictive control. *Automatica*, 47 (3), 571–577, doi: 10.1016/j.automatica.2011.01.009. URL http://www.sciencedirect.com/science/article/pii/S0005109811000240.

18 Bayat, F., Johansen, T.A., and Jalali, A.A. (2012) Flexible piecewise function evaluation methods based on truncated binary search trees and lattice representation in explicit MPC. *IEEE Transactions on Control Systems Technology*, 20 (3), 632–640, doi: 10.1109/TCST.2011.2141134.

19 Airan, A., Bhartiya, S., and Bhushan, M. (2013) Linear machine: a novel approach to point location problem, in *10th IFAC International Symposium on Dynamics and Control of Process Systems (2013)*, International Federation of Automatic Control, pp. 445–450.

20 Herceg, M., Mariéthoz, S., and Moraria, M. (2013) Evaluation of piecewise affine control law via graph traversal, in *European Control Conference*, Zurich, Switzerland, pp. 3083–3088. URL http://control.ee.ethz.ch/index.cgi?page=publications;action=details;id=4439.

21 Jafargholi, M., Peyrl, H., Zanarini, A., Herceg, M., and Mariéthoz, S. (2014) Accelerating space traversal methods for explicit model predictive control via space partitioning trees, in *European Control Conference*, Strasbourg, France, pp. 103–108. URL http://control.ee.ethz.ch/index.cgi?page=publications;action=details;id=4956.

22 Martinez-Rodriguez, M.C., Brox, P., and Baturone, I. (2015) Digital VLSI implementation of Piecewise-Affine controllers based on lattice approach. *IEEE Transactions on Control Systems Technology*, 23 (3), 842–854, doi: 10.1109/TCST.2014.2345094.

23 Oliveri, A., Gianoglio, C., Ragusa, E., and Storace, M. (2015) Low-complexity digital architecture for solving the point location problem in explicit model predictive control. *Journal of the Franklin Institute*, 352 (6), 2249–2258, doi: 10.1016/j.jfranklin.2015.03.018. URL http://www.sciencedirect.com/science/article/pii/S0016003215001167.

24 Zhang, J., Xiu, X., Xie, Z., and Hu, B. (2015) Using a two-level structure to manage the point location problem in explicit model predictive control. *Asian Journal of Control*, 18 (3), 1075–1086, doi: 10.1002/asjc.1178. URL http://dx.doi.org/10.1002/asjc.1178.

25 Airan, A., Bhushan, M., and Bhartiya, S. (2016) Linear machine solution to point location problem. *IEEE Transactions on Automatic Control*, 62 (3), 1403–1410, doi: 10.1109/TAC.2016.2573201.

11

Extensions to Other Classes of Problems

The approach described previously in Chapter 10 is not limited to deterministic model predictive control (MPC) problems. As a matter of fact, it can be applied in a variety of problem formulations, including, among others, reference and trajectory tracking, MPC under measurable uncertainty, hybrid MPC, and moving horizon estimation. Brief overviews are presented and discussed in this chapter.

11.1 Hybrid Explicit MPC

Discrete decisions are oftentimes incorporated within MPC rolling horizon formulations. Thereby, the MPC formulation needs to make decisions that are modeled via either binary or integer decision variables. In the case of integer variables, we assume that the reformulation described in [1] can take place to replace the integer variables with sums of binary variables. Therefore, for the remaining of the section, we will only consider binary variables. Several occurrences of binary variables may be present within MPC formulations and the underlying linear discrete time, time invariant state-space models. Consider, for example, the cases described in Eqs. (11.1) and (11.2) and their combinations. Note that the binary variable dependent bounds of inputs, states, and outputs need to be linear (or exactly linearizable) with respect to the binary variables to preserve the quadratic and convex functional form of the resulting MPC formulation. The model

Multi-parametric Optimization and Control, First Edition.
Efstratios N. Pistikopoulos, Nikolaos A. Diangelakis, and Richard Oberdieck.

states and outputs are dependent linearly on a variable vector b_k, which corresponds to the binary variables.

$$\begin{cases} x_{k+1} = Ax_k + B\left[u_k^T, b_k^T\right]^T, \\ y_k = Cx_k + D\left[u_k^T, b_k^T,\right]^T \\ u_{min} \leq u_k \leq u_{max}, \quad \forall k \in [0, N-1], \\ x_{min} \leq x_k \leq x_{max}, \quad \forall k \in [0, N-1], \\ y_{min} \leq y_k \leq y_{max}, \quad \forall k \in [0, N-1], \\ b_k \in \{0, 1\}, \quad \forall k \in [0, N-1], \end{cases} \tag{11.1}$$

$$\begin{cases} x_{k+1} = Ax_k + Bu_k, \\ y_k = Cx_k + Du_k, \\ u_{min}(b_k) \leq u_k \leq u_{max}(b_k), \quad \forall k \in [0, N-1], \\ x_{min}(b_k) \leq x_k \leq x_{max}(b_k), \quad \forall k \in [0, N], \\ y_{min}(b_k) \leq y_k \leq y_{max}(b_k), \quad \forall k \in [0, N-1], \\ b_k \in \{0, 1\}, \quad \forall k \in [0, N]. \end{cases} \tag{11.2}$$

There are a few things to be noted regarding the handling of binary variables in multi-parametric model predictive control (mp-MPC):

- The binary variable vector can undergo the same reformulations as shown before for the continuous variables.
- Depending on the formulation of the problem, bilinear and/or quadratic binary terms may occur in the objective function. In this case POP can handle the forms automatically as long as the resulting multi-parametric mixed-integer quadratic programming (mp-MIQP) is of convex functional form (see Chapter 8). POP can currently handle binary variables in the constraints only in linear form.
- On the contrary to (mp-)MPC problems involving only continuous variables, in the case of hybrid mp-MPC, a solver that can handle mixed integer quadratic and/or linear multi-parametric programming problems is needed.
- Hybrid mp-MPC problems involving state space formulations and constraints of the form of (11.1) and (11.2) are guaranteed to produce mp-MIQP's of convex functional forms.

Since the problem remains linear in the state-space definition and its bounds, then the linear quadratic regulator (LQR) formulation corresponds to an mp-MIQP problem. Equivalently, a linear MPC cost function would yield an mp-MILP problem, as previously discussed. Assuming that the output and control horizons are equal, the state vector at every time step is expressed via Eq. (11.3) and/or (11.3).

$$\begin{aligned} x_k &= A^k x_0 + \sum_{i=0}^{k-1} A^{k-i-1} B[u_i^T, b_i^T]^T, \\ &\forall k \in [1, N], u_{min} \leq u_k \leq u_{max}, \\ &\forall k \in [0, N-1], x_{min} \leq x_k \leq x_{max}, \\ &\forall k \in [0, N-1], y_{min} \leq y_k \leq y_{max}, \\ &\forall k \in [0, N-1], b_k \in \{0, 1\}, \forall k \in [0, N-1], \end{aligned} \tag{11.3}$$

$$\begin{aligned} x_k &= A^k x_0 + \sum_{i=0}^{k-1} A^{k-i-1} B u_i, \\ &\forall k \in [1, N], u_{min}(b_k) \leq u_k \leq u_{max}(b_k), \\ &\forall k \in [0, N-1], x_{min}(b_k) \leq x_k \leq x_{max}(b_k), \\ &\forall k \in [0, N], y_{min}(b_k) \leq y_k \leq y_{max}(b_k), \\ &\forall k \in [0, N-1], b_k \in \{0, 1\},\ \forall k \in [0, N]. \end{aligned} \tag{11.4}$$

Therefore, the state vector at every step of the output and control horizon can be expressed as a linear function of the continuous and binary variables. The rest of the multi-parametric quadratic programming (mp-QP) formulation steps can be followed and the resulting mp-QP would be of the form of Eq. (11.5):

$$\begin{aligned} \underset{u}{\text{Minimize}}\ & \omega^T H \omega + \omega^T Z x_0 \\ \text{Subject to}\ & G\omega \leq W + S x_0 \\ & \text{CR}_A x_0 \leq \text{CR}_b \\ & \omega = [u^T, b^T]^T. \end{aligned} \tag{11.5}$$

11.1.1 Explicit Hybrid MPC – An Example of mp-MPC

In this section we will consider the discrete time, linear, time invariant state-space model described in Eq. (10.25), and similar control design as before. On the contrary to the previous example, the bounds of the control actions now include the option to range between −1.5 and 1.5. This option is represented as a binary decision variable b_k. Additionally to the cost associated with the continuous value of the control variables in the objective function, the possibility to apply an action more than 1 or less than −1 is further penalized. The control and output horizons will be decreased to 5 as the computational expense of the problem is expected to become significant.

$$A = \begin{bmatrix} 1 & 1 \\ 0 & 1 \end{bmatrix},\ B = \begin{bmatrix} 0 \\ 1 \end{bmatrix},\ C = \begin{bmatrix} 1 & 2 \end{bmatrix},\ D = \begin{bmatrix} 0. \end{bmatrix} \tag{11.6}$$

$$\begin{aligned} &\text{Bounds on states:} && \begin{bmatrix} -10 \\ -10 \end{bmatrix} \le x_k \le \begin{bmatrix} 10 \\ 10 \end{bmatrix}, \forall k \in [0, N], \\ &\text{Bounds on inputs:} && -1 - 0.5b_k \le u_k \le 1 + 0.5b_k, \\ &&& \forall k \in [0, N-1], \\ &\text{Bounds on outputs:} && -25 \le y_k \le 25, \forall k \in [0, N]. \end{aligned} \tag{11.7}$$

$$Q = \begin{bmatrix} 1 & 0 \\ 0 & 1 \end{bmatrix},\ R = 0.01, \tag{11.8}$$

$$\begin{aligned} &\text{Prediction horizon: } N = 5, \\ &\text{Control horizon: } \quad M = 5. \end{aligned} \tag{11.9}$$

$$A_T = \begin{bmatrix} 0.6136 & 1.6099 \\ -0.3742 & -0.3682 \\ -0.6136 & -1.6099 \\ 0.3742 & 0.3682 \end{bmatrix},\ b_T = \begin{bmatrix} 1 \\ 1 \\ 1 \\ 1 \end{bmatrix}. \tag{11.10}$$

The overall hybrid mp-MPC problem formulation is presented in Eq. (11.11). Note the linear term associated with the binary decisions in the objective function. This term corresponds to the penalization of augmenting the control variable bounds.

$$\begin{aligned} \underset{u,b}{\text{Minimize}}\ & x_5^T \begin{bmatrix} 2.6235 & 1.6296 \\ 1.6296 & 2.6457 \end{bmatrix} x_5 + \sum_{k=0}^{4} x_k^T \begin{bmatrix} 1 & 0 \\ 0 & 1 \end{bmatrix} x_k \\ & + u_k^T 0.01 u_k + 1000 b_k \\ \text{Subject to}\ & \begin{cases} x_{k+1} = \begin{bmatrix} 1 & 1 \\ 0 & 1 \end{bmatrix} x_k + \begin{bmatrix} 0 \\ 1 \end{bmatrix} u_k, & \forall k \in [0, 4] \\ y_k = \begin{bmatrix} 1 & 2 \end{bmatrix} x_k \end{cases} \\ & \begin{bmatrix} -10 \\ -10 \end{bmatrix} \le x_k \le \begin{bmatrix} 10 \\ 10 \end{bmatrix}, \quad \forall k \in [0, 5] \end{aligned}$$

$$
\begin{aligned}
& -25 \leq y_k \leq 25, \quad \forall k \in [0,5] \\
& -1 - 0.5b_k \leq u_k \leq 1 + 0.5b_k, \quad \forall k \in [0,4] \\
& \begin{bmatrix} 0.6136 & 1.6099 \\ -0.3742 & -0.3682 \\ -0.6136 & -1.6099 \\ 0.3742 & 0.3682 \end{bmatrix} x_5 \leq \begin{bmatrix} 1 \\ 1 \\ 1 \\ 1 \end{bmatrix} \\
& b_k \in \{0,1\}, \quad \forall k \in [0,4].
\end{aligned} \tag{11.11}
$$

The reformulated problem corresponds to an mp-MIQP where the states are treated as unknown but bounded parameters. The reformulated problem features vectors of 5 continuous and 5 binary variables, u and b, respectively, and a total of 32 constraints. The solution of the problem yields 3167 critical regions using the connected graph algorithm and the exact comparison procedure described in [2] and discussed in Chapter 6.

11.1.2 Results and Validation

The solution of the mp-MIQP problem corresponding to the hybrid controller of Eq. (11.11) is presented in Figure 11.1 and Table 11.1. More specifically, Figure 11.1 presents the 3167 critical regions. The analytical solution of 2 critical regions is presented in Table 11.1. It is noteworthy that the overall feasible space of the hybrid mp-MPC is identical to the feasible space of an mp-MPC solution where the bounds of the system are set to $-1.5 \leq u_k \leq 1.5, \ \forall k \in [0,4]$ (hereon referred to as "extended mp-MPC"). The optimal trajectory of the continuous inputs, though, differs. The latter is a direct result of the extra cost of switching in the objective function. As soon as the state trajectory enters a space where an action bounded by $-1 \leq u_k \leq 1, \ \forall k \in [0,4]$ (hereon referred to as "regular mp-MPC") is feasible and favorable, the costly binary variables are switched off, thereby affecting positively the minimization of the cost function of the problem. This is shown through a comparative simulation between the extended mp-MPC, the hybrid mp-MPC, and the regular mp-MPC. Note that the simulation of the feasible spaces as a function of the states of the regular and hybrid mp-MPC may differ. This is shown in Figure 11.2.

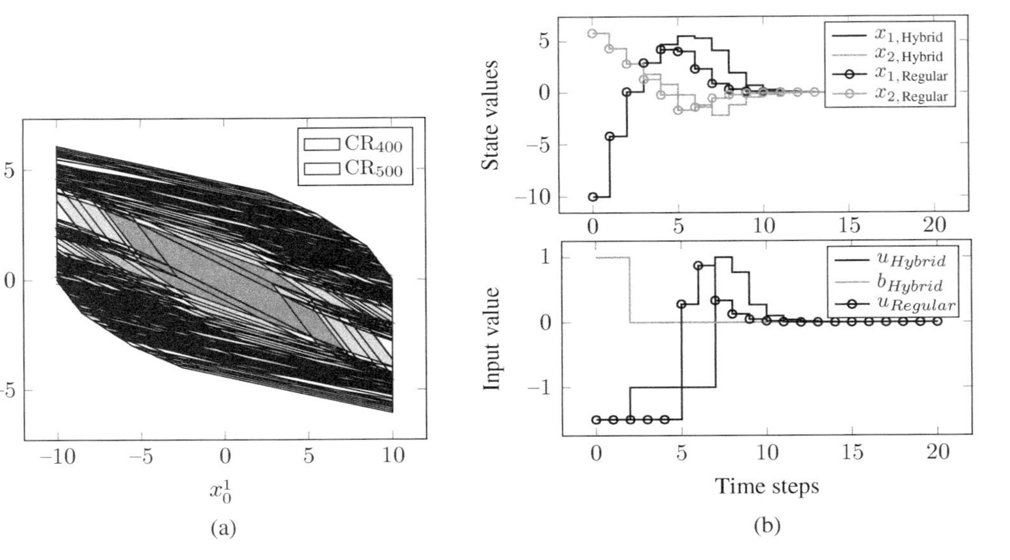

Figure 11.1 The solution of the mp-MPC example problem. Left: (a) The partitioning of the parameter space into 3100 critical regions is shown, each of which is associated with a different set of active constraints k_i. The legend shows only 2 out of the 3100 critical regions to simplify the figure. The full solutions to the same regions is presented in Table 11.1. Right: (b) The linear, time invariant, state-space, model is simulated in a closed loop fashion. In this particular snapshot, the initial state values have been chosen to be $x_0^1 = -10$ and $x_0^2 = 5.8$. Observe how the controller is able to stabilize the system around the origin and achieve a final value for the state $x_f^1 = 0$ and $x_f^2 = 0$ while determining both the continuous and discrete decisions.

Table 11.1 The partial analytical solution of the mp-MPC problem.

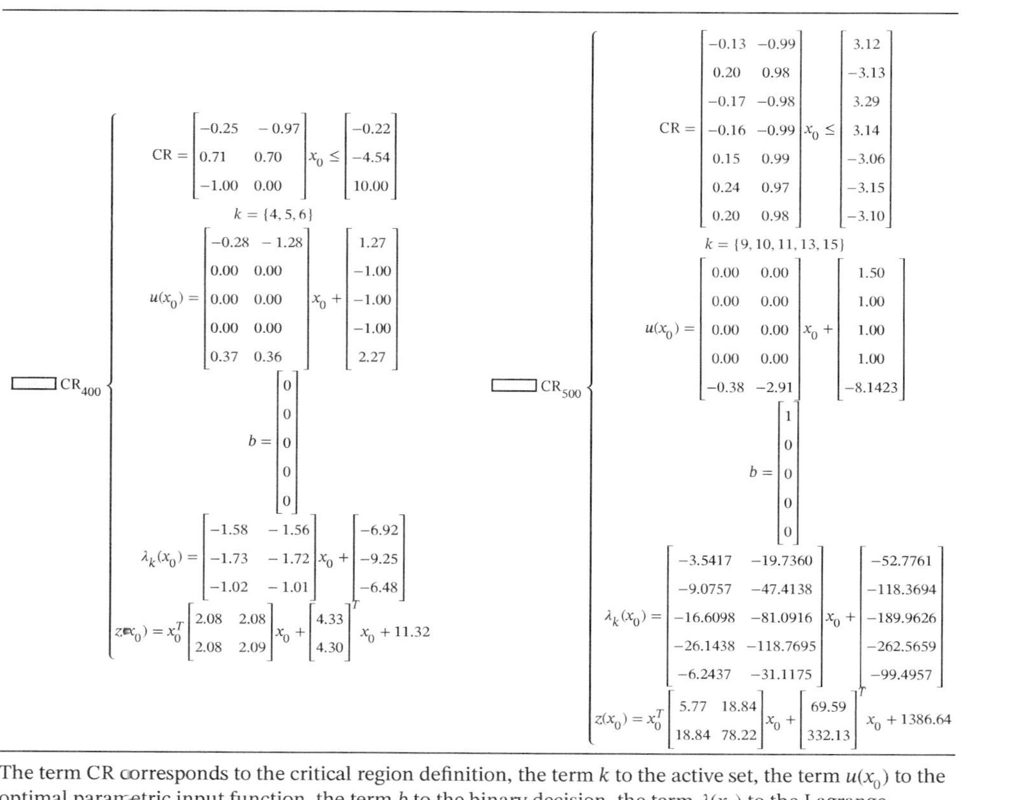

$$
\boxed{}\ \mathrm{CR}_{400}
\begin{cases}
\mathrm{CR} = \begin{bmatrix} -0.25 & -0.97 \\ 0.71 & 0.70 \\ -1.00 & 0.00 \end{bmatrix} x_0 \leq \begin{bmatrix} -0.22 \\ -4.54 \\ 10.00 \end{bmatrix} \\
k = \{4, 5, 6\} \\
u(x_0) = \begin{bmatrix} -0.28 & -1.28 \\ 0.00 & 0.00 \\ 0.00 & 0.00 \\ 0.00 & 0.00 \\ 0.37 & 0.36 \end{bmatrix} x_0 + \begin{bmatrix} 1.27 \\ -1.00 \\ -1.00 \\ -1.00 \\ 2.27 \end{bmatrix} \\
b = \begin{bmatrix} 0 \\ 0 \\ 0 \\ 0 \\ 0 \end{bmatrix} \\
\lambda_k(x_0) = \begin{bmatrix} -1.58 & -1.56 \\ -1.73 & -1.72 \\ -1.02 & -1.01 \end{bmatrix} x_0 + \begin{bmatrix} -6.92 \\ -9.25 \\ -6.48 \end{bmatrix} \\
z(x_0) = x_0^T \begin{bmatrix} 2.08 & 2.08 \\ 2.08 & 2.09 \end{bmatrix} x_0 + \begin{bmatrix} 4.33 \\ 4.30 \end{bmatrix}^T x_0 + 11.32
\end{cases}
$$

$$
\boxed{}\ \mathrm{CR}_{500}
\begin{cases}
\mathrm{CR} = \begin{bmatrix} -0.13 & -0.99 \\ 0.20 & 0.98 \\ -0.17 & -0.98 \\ -0.16 & -0.99 \\ 0.15 & 0.99 \\ 0.24 & 0.97 \\ 0.20 & 0.98 \end{bmatrix} x_0 \leq \begin{bmatrix} 3.12 \\ -3.13 \\ 3.29 \\ 3.14 \\ -3.06 \\ -3.15 \\ -3.10 \end{bmatrix} \\
k = \{9, 10, 11, 13, 15\} \\
u(x_0) = \begin{bmatrix} 0.00 & 0.00 \\ 0.00 & 0.00 \\ 0.00 & 0.00 \\ 0.00 & 0.00 \\ -0.38 & -2.91 \end{bmatrix} x_0 + \begin{bmatrix} 1.50 \\ 1.00 \\ 1.00 \\ 1.00 \\ -8.1423 \end{bmatrix} \\
b = \begin{bmatrix} 1 \\ 0 \\ 0 \\ 0 \\ 0 \end{bmatrix} \\
\lambda_k(x_0) = \begin{bmatrix} -3.5417 & -19.7360 \\ -9.0757 & -47.4138 \\ -16.6098 & -81.0916 \\ -26.1438 & -118.7695 \\ -6.2437 & -31.1175 \end{bmatrix} x_0 + \begin{bmatrix} -52.7761 \\ -118.3694 \\ -189.9626 \\ -262.5659 \\ -99.4957 \end{bmatrix} \\
z(x_0) = x_0^T \begin{bmatrix} 5.77 & 18.84 \\ 18.84 & 78.22 \end{bmatrix} x_0 + \begin{bmatrix} 69.59 \\ 332.13 \end{bmatrix}^T x_0 + 1386.64
\end{cases}
$$

The term CR corresponds to the critical region definition, the term k to the active set, the term $u(x_0)$ to the optimal parametric input function, the term b to the binary decision, the term $\lambda(x_0)$ to the Lagrange multipliers of the active constraints, and the term $z(x_0)$ to the parametric objective function.

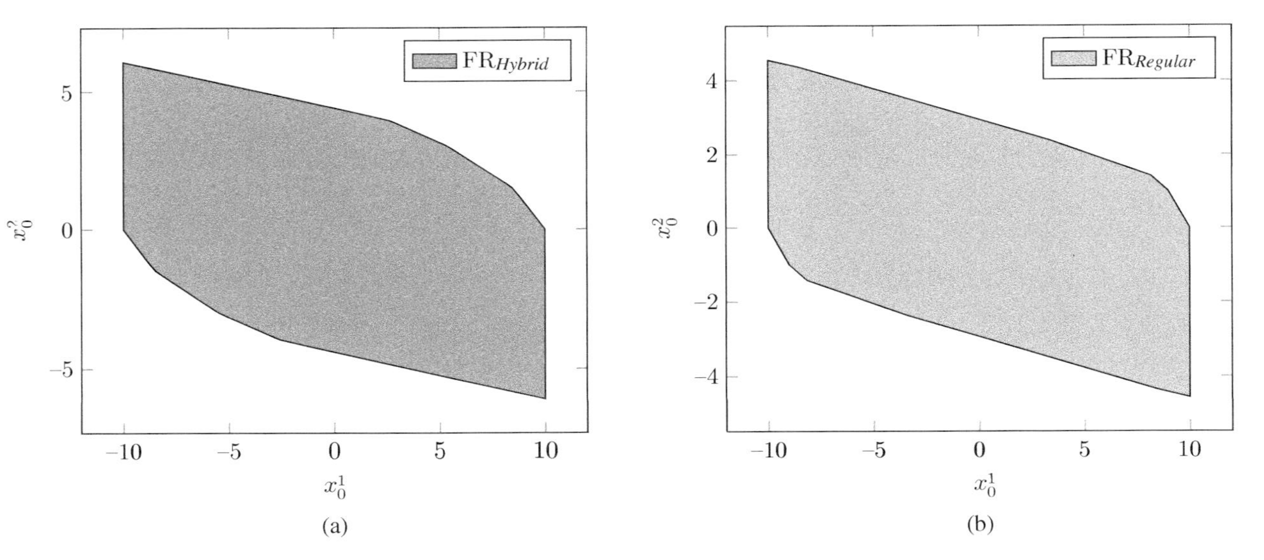

Figure 11.2 Comparison of the feasible regions between (a) the hybrid mp-MPC and (b) the regular mp-MPC.

11.2 Disturbance Rejection

In many instances, MPC rolling horizon formulations are used to reject measurable and/or unmeasurable disturbances that may occur to a process/system. These disturbances are typically system variables over which the user has no influence. Those disturbances can only be handled by adjusting the control actions in such a way that they affect the outcome of the optimization problem as little as possible. Here we briefly discuss the effect on measurable disturbances that have a linear correlation to the system states and/or outputs.

Consider, for example, the discrete time, linear, time invariant state-space model described in Eq. (11.12). The model states and outputs are dependent linearly on a variable vector d_k, which corresponds to the measured disturbance vector.

$$\begin{cases} x_{k+1} = Ax_k + Bu_k + \tilde{B}d_k, \\ y_k = Cx_k + Du_k + \tilde{D}d_k. \end{cases} \tag{11.12}$$

There are a few things to be noted regarding the handling of measurable disturbances in mp-MPC:

- The disturbance vector is treated as a source of uncertainty within the mp-QP, therefore, it must be bounded.
- On the contrary to the state vector, where only the initial value of the state vector is uncertain, the disturbance vector is uncertain throughout the horizon considered in the MPC problem. Therefore, the next bullet point naturally follows.
- A decision needs to be made regarding the length of the uncertainty considered within the problem. This decision highly depends on the nature of the problem at hand, specifically knowledge of uncertainty propagation. Typically, a single step is considered. This decision is twofold. On one hand, the parametric vector of the problem is only augmented by the size of the disturbance vector and on the other, no assumptions regarding the propagation of the uncertainty into the future are necessary.

Since the problem remains linear, then the LQR formulation can still correspond to an mp-QP problem. Equivalently, a linear MPC cost function would yield an mp-LP problem, as previously discussed. Assuming that only a single step into the future is considered in terms of uncertainty realization (i.e. $d_k = d \; \forall k \in [1, N]$) and that the output and control horizons are equal the state vector at every time step is expressed via Eq. (11.13).

$$x_k = A^k x_0 + \sum_{i=0}^{k-1} A^{k-i-1} B u_i + \sum_{i=0}^{k-1} A^{k-i-1} \tilde{B} d, \quad \forall k \in [1, N]. \tag{11.13}$$

Therefore, the equivalent of Eq. (10.21) can be derived (Eq. (11.14)). The term A_r^k now also includes the linear relation of the kth state to the disturbance. The rest of the mp-QP formulation steps can be followed and the resulting mp-QP would be of the form of Eq. (11.15):

$$\begin{aligned} x_k &= A_r^k \left[x_0^T \; d^T\right]^T + B_r^k u, \\ u &= [u_1^T, u_2^T, \ldots, u_{k-1}^T]^T, \end{aligned} \tag{11.14}$$

$$\begin{aligned} &\underset{u}{\text{Minimize}} \; u^T H u + u^T Z \left[x_0^T \; d^T\right]^T \\ &\text{Subject to } Gu \leq W + S\left[x_0^T \; d^T\right]^T \\ &\qquad\qquad \text{CR}_A \left[x_0^T \; d^T\right]^T \leq \text{CR}_b. \end{aligned} \tag{11.15}$$

11.2.1 Explicit Disturbance Rejection – An Example of mp-MPC

Consider the discrete-time, linear, time invariant model of Eq. (11.12). In this example, we consider the following system dynamics (Eq. (11.16)), box constraints (Eq. (11.17)), objective function weights (Eq. (11.18)), horizons (Eq. (11.19)), and terminal constraints (Eq. (11.20)).

$$A = \begin{bmatrix} 1 & 1 \\ 0 & 1 \end{bmatrix}, \; B = \begin{bmatrix} 0 \\ 1 \end{bmatrix}, \; \tilde{B} = \begin{bmatrix} -1 \\ 1 \end{bmatrix}, \; C = \begin{bmatrix} 1 & 2 \end{bmatrix}, \; D = \begin{bmatrix} 0. \end{bmatrix} \tag{11.16}$$

$$\begin{aligned} &\text{Bounds on states:} && \begin{bmatrix} -10 \\ -10 \end{bmatrix} \leq x_k \leq \begin{bmatrix} 10 \\ 10 \end{bmatrix}, && \forall k \in [0, N], \\ &\text{Bounds on inputs:} && -1 \leq u_k \leq 1, && \forall k \in [0, N-1], \\ &\text{Bounds on outputs:} && -25 \leq y_k \leq 25, && \forall k \in [0, N], \end{aligned}$$

$$\text{Bounds on disturbance: } -0.5 \leq d = d_k \leq 0.5,\ \forall k \in [0, N]. \tag{11.17}$$

$$Q = \begin{bmatrix} 1 & 0 \\ 0 & 1 \end{bmatrix},\ R = 0.01. \tag{11.18}$$

$$\begin{aligned} &\text{Prediction horizon: } N = 10, \\ &\text{Control horizon: } \quad M = 10. \end{aligned} \tag{11.19}$$

$$A_T = \begin{bmatrix} 0.3358 & 0.8809 & -0.3336 \\ -0.2553 & -0.2512 & 0.9336 \\ -0.0001 & -0.0001 & 1 \\ 0.0001 & 0.0001 & -1 \\ 0.2553 & 0.2512 & -0.9336 \\ -0.3358 & -0.8809 & 0.3336 \end{bmatrix},\ b_T = \begin{bmatrix} 0.5472 \\ 0.6823 \\ 0.6961 \\ 0.6961 \\ 0.6823 \\ 0.5472 \end{bmatrix}. \tag{11.20}$$

Given Eqs. (10.25)–(10.29) and the MPC formulation in Eq. (10.18) the MPC problem are presented in Eq. (11.21).

$$\begin{aligned} \underset{u}{\text{Minimize}}\ & x_{10}^T \begin{bmatrix} 2.6235 & 1.6296 \\ 1.6296 & 2.6457 \end{bmatrix} x_{10} + \sum_{k=0}^{9} x_k^T \begin{bmatrix} 1 & 0 \\ 0 & 1 \end{bmatrix} x_k \\ & + u_k^T 0.01 u_k \\ \text{Subject to}\ & \begin{cases} x_{k+1} = \begin{bmatrix} 1 & 1 \\ 0 & 1 \end{bmatrix} x_k + \begin{bmatrix} 0 \\ 1 \end{bmatrix} u_k + \begin{bmatrix} -1 \\ 1 \end{bmatrix} d_k, \ \forall k \in [0, 9] \\ y_k = \begin{bmatrix} 1 & 2 \end{bmatrix} x_k \end{cases} \\ & \begin{bmatrix} -10 \\ -10 \end{bmatrix} \leq x_k \leq \begin{bmatrix} 10 \\ 10 \end{bmatrix}, \quad \forall k \in [0, 10] \\ & -25 \leq y_k \leq 25, \quad \forall k \in [0, 10] \\ & -1 \leq u_k \leq 1, \quad \forall k \in [0, 9] \\ & -0.5 \leq d_k = d \leq 0.5, \quad \forall k \in [0, 9] \\ & \begin{bmatrix} 0.3358 & 0.8809 & -0.3336 \\ -0.2553 & -0.2512 & 0.9336 \\ -0.0001 & -0.0001 & 1 \\ 0.0001 & 0.0001 & -1 \\ 0.2553 & 0.2512 & -0.9336 \\ -0.3358 & -0.8809 & 0.3336 \end{bmatrix} \begin{bmatrix} x_{10} \\ d \end{bmatrix} \leq \begin{bmatrix} 0.5472 \\ 0.6823 \\ 0.6961 \\ 0.6961 \\ 0.6823 \\ 0.5472 \end{bmatrix}. \end{aligned} \tag{11.21}$$

11.2.2 Results and Validation

The reformulation of the MPC and solution to the mp-QP of Eq. (11.21) yields a solution of 189 critical regions using the connected graph algorithm. Every critical region has a distinct parametric optimal function set for the 10 optimization variables u. Treating the disturbance as a parameter results into a three-dimensional parametric space. Therefore, instead of providing 3D graphs of the critical regions, which would be ineligible, we provide the projected critical regions on the x_0^1, d and x_0^2, d space, while fixing the remaining parameter to 0. The projection on the x_0^1, x_0^2 for a disturbance value of $d = 0$ is, as expected, identical to the one presented in Figure 10.2. For reference, we provide the full solution of one critical region to show the disturbance dependence of the optimal action and CR (Table 11.2; Figure 11.3).

Furthermore, we show the overall feasible parametric space of the problem solution in Figure 11.4a as well as the results of the closed loop simulation in (b). Regarding the latter, the controller regulates the system to its origin under disturbance. The 2D feasible space presented here is a projection on the x_0^1–x_0^2 plane, i.e. the gray area corresponds to the maximal set of feasible values for the system states.

11.3 Reference Trajectory Tracking

Reference and trajectory tracking is a common problem for which receding horizon MPC policies are applied. Similarly to the LQR case shown in Chapter 10, this problem can also be reformulated to correspond to an mp-QP problem. In this section we show such a reformulation, solution, and closed-loop validation. We also consider path constraints within the problem formulation that must be satisfied on the discrete system. In this class of problems, the objective function may correspond to a penalization of an output deviation to a reference trajectory. In the context of multi-parametric programming, the reference is treated as an extra parameter in the problem formulation. Furthermore, in order to stabilize the system around a trajectory, consecutive input deviation terms may be incorporated. The problem MPC formulation considered here is presented in Eq. (11.22).

$$
\begin{aligned}
&\underset{u}{\text{Minimize}} \ \sum_{k=0}^{N} S_k^T \mathrm{QR}(k) S_k + \sum_{k=0}^{M-1} \Delta u_k^T R1 \Delta u_k \\
&\text{Subject to} \begin{cases} x_{k+1} = Ax_k + Bu_k \\ y_k = Cx_k \end{cases} && \forall k \in [0, M-1] \\
&\qquad\qquad\ \begin{cases} x_{k+1} = Ax_k + Bu_{M-1} \\ y_k = Cx_k \end{cases} && \forall k \in [M, N-1] \\
&\qquad\qquad\ \Delta u_k = u_k - u_{k-1}, && \forall k \in [0, M-1] \\
&\qquad\qquad\ S_k = y_i^{ref} - y_{i,k}, && \forall k \in [0, N] \\
&\qquad\qquad\ x_{\min} \le x_k \le x_{\max}, && \forall k \in [0, N] \\
&\qquad\qquad\ y_{\min} \le y_k \le y_{\max}, && \forall k \in [0, N] \\
&\qquad\qquad\ u_{\min} \le u_k \le u_{\max}, && \forall k \in [0, M-1] \\
&\qquad\qquad\ A_T x_N \le b_T.
\end{aligned}
\tag{11.22}
$$

Table 11.2 The solution of a CR of the mp-MPC problem with disturbance.

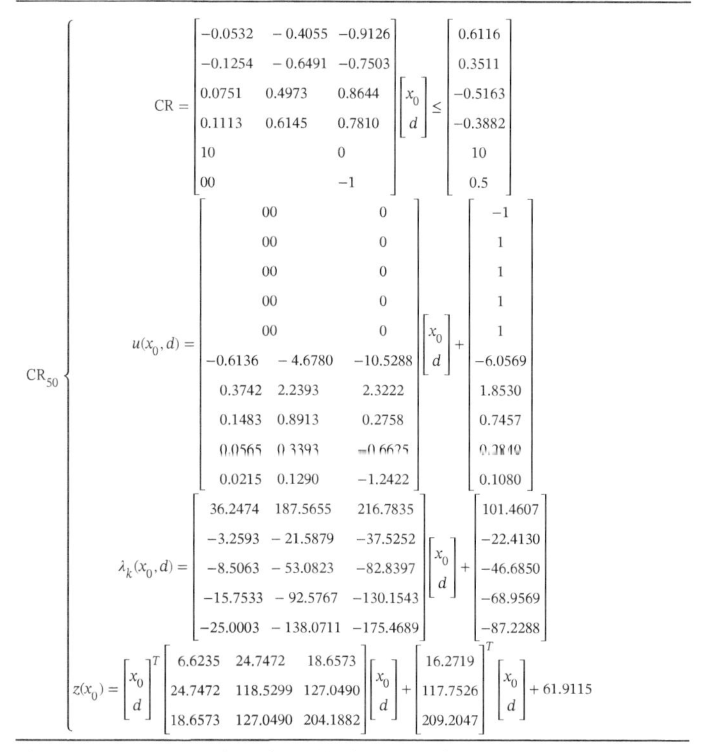

$$
\mathrm{CR}_{50}
\begin{cases}
\mathrm{CR} = \begin{bmatrix} -0.0532 & -0.4055 & -0.9126 \\ -0.1254 & -0.6491 & -0.7503 \\ 0.0751 & 0.4973 & 0.8644 \\ 0.1113 & 0.6145 & 0.7810 \\ 10 & & 0 \\ 00 & & -1 \end{bmatrix} \begin{bmatrix} x_0 \\ d \end{bmatrix} \le \begin{bmatrix} 0.6116 \\ 0.3511 \\ -0.5163 \\ -0.3882 \\ 10 \\ 0.5 \end{bmatrix} \\[2ex]
u(x_0, d) = \begin{bmatrix} 00 & 0 \\ 00 & 0 \\ 00 & 0 \\ 00 & 0 \\ 00 & 0 \\ -0.6136 \quad -4.6780 & -10.5288 \\ 0.3742 \quad 2.2393 & 2.3222 \\ 0.1483 \quad 0.8913 & 0.2758 \\ 0.0565 \quad 0.3393 & -0.6625 \\ 0.0215 \quad 0.1290 & -1.2422 \end{bmatrix} \begin{bmatrix} x_0 \\ d \end{bmatrix} + \begin{bmatrix} -1 \\ 1 \\ 1 \\ 1 \\ 1 \\ -6.0569 \\ 1.8530 \\ 0.7457 \\ 0.2849 \\ 0.1080 \end{bmatrix} \\[2ex]
\lambda_k(x_0, d) = \begin{bmatrix} 36.2474 & 187.5655 & 216.7835 \\ -3.2593 & -21.5879 & -37.5252 \\ -8.5063 & -53.0823 & -82.8397 \\ -15.7533 & -92.5767 & -130.1543 \\ -25.0003 & -138.0711 & -175.4689 \end{bmatrix} \begin{bmatrix} x_0 \\ d \end{bmatrix} + \begin{bmatrix} 101.4607 \\ -22.4130 \\ -46.6850 \\ -68.9569 \\ -87.2288 \end{bmatrix} \\[2ex]
z(x_0) = \begin{bmatrix} x_0 \\ d \end{bmatrix}^T \begin{bmatrix} 6.6235 & 24.7472 & 18.6573 \\ 24.7472 & 118.5299 & 127.0490 \\ 18.6573 & 127.0490 & 204.1882 \end{bmatrix} \begin{bmatrix} x_0 \\ d \end{bmatrix} + \begin{bmatrix} 16.2719 \\ 117.7526 \\ 209.2047 \end{bmatrix}^T \begin{bmatrix} x_0 \\ d \end{bmatrix} + 61.9115
\end{cases}
$$

The term CR corresponds to the critical region definition, the term $u(x_0, d)$ to the optimal parametric input function, the term $\lambda(x_0, d)$ to the Lagrange multipliers of the active constraints, and the term $z(x_0, d)$ to the parametric objective function.

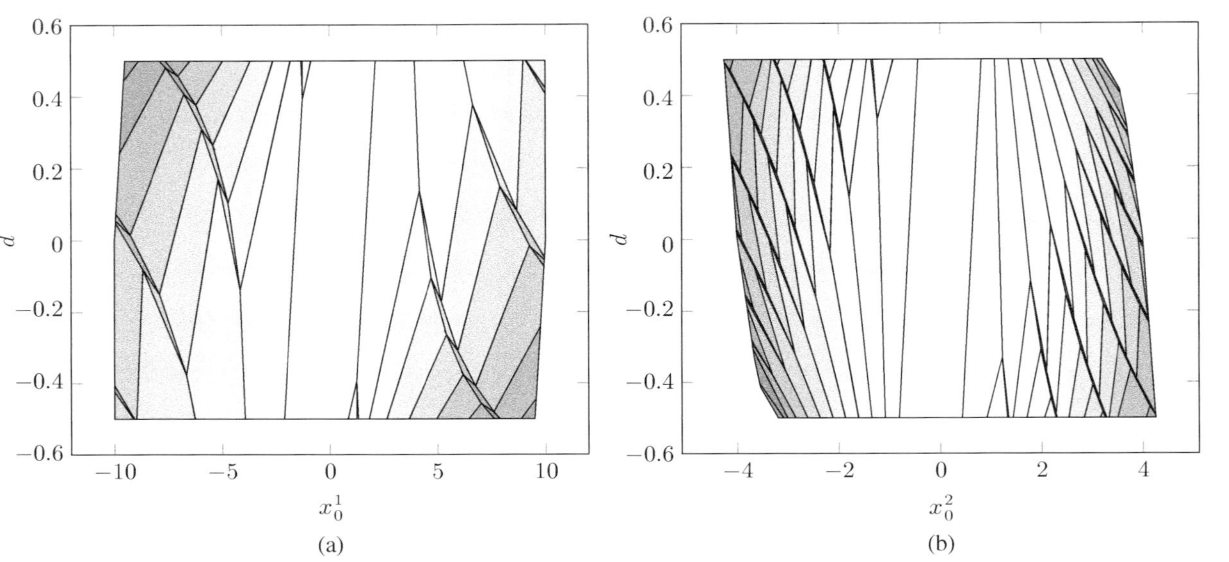

Figure 11.3 The solution of the mp-MPC with disturbance example problem. (a) The partitioning of the parameter space into 53 critical regions is shown, each of which is associated with a different set of active constraints k_i. The legend shows only 2 out of the 53 critical regions to simplify the figure. This is a 2D projection on the x_0^1–d space, therefore the total 189 solutions are reduced when a parameter is fixed for plotting purposes, in this case $x_0^2 = 0$. (b) The partitioning of the parameter space into 127 critical regions is shown, each of which is associated with a different set of active constraints k_i. The legend shows only 2 out of the 127 critical regions to simplify the figure. This is a 2D projection on the x_0^2–d space, therefore the total 189 solutions are reduced when a parameter is fixed for plotting purposes, in this case $x_0^1 = 0$.

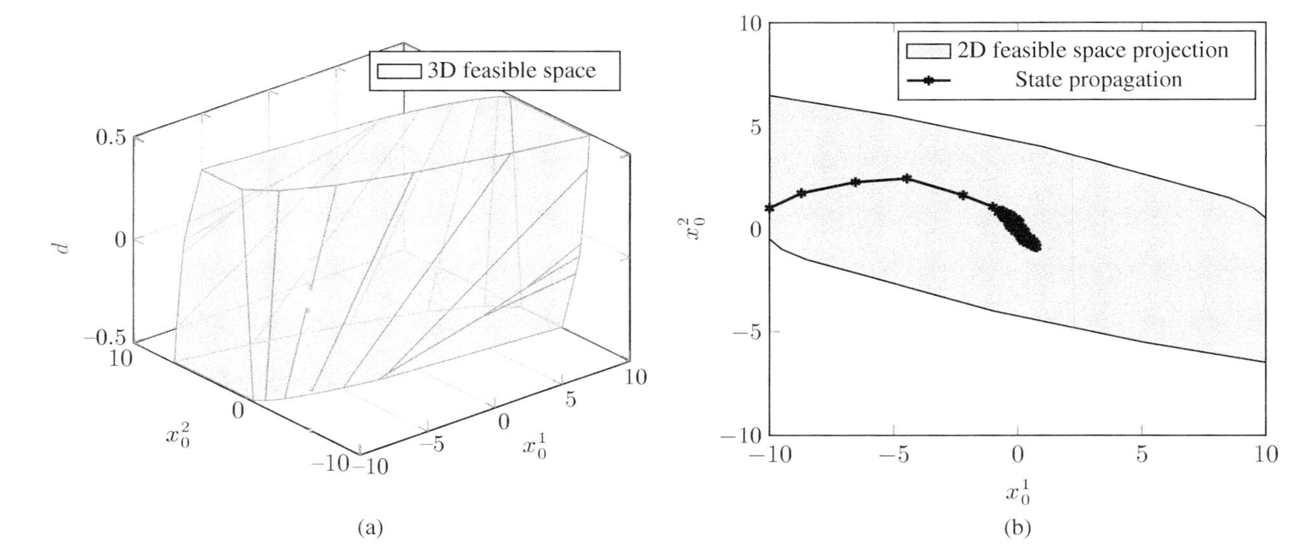

Figure 11.4 The solution of the mp-MPC with disturbance example problem. (a) The three dimensional representation of the feasible space of the mp-MPC with disturbance solution. This is the concatenated space of 189 critical regions on the states and disturbance space. (b) The two dimensional projection of the feasible space in (a) along with an example closed-loop simulation of the system in Eq. (11.21), starting from $x_0^1 = -10$, $x_0^1 = -1$ and random feasible values for the disturbance d.

In Eq. (11.22) y_i^{ref} corresponds to the reference value for the ith tracked output y_i. Note that the reference value is treated as a parameter and assumed fixed throughout the prediction and control horizon in order to reduce the number of parameters of the problem. An alternative formulation could incorporate $y_i^{ref}(k), \forall k \in [0, N]$, in which case $i \times N + 1$ parameters associated with the reference tracking would be introduced. The term u_{k-1} corresponds to the previous optimal control action. At $k = 0$ the previous control action is known during the application of the solution but unknown during the solution of the problem. Therefore u_{-1} is treated as a parameter vector. To summarize, on the contrary to the LQR problem, here the initial state values (x_0), the reference values (y_i^{ref} or $y_i^{ref}(k), \forall k \in [0, N]$), and the previous control action at the initial time step (u_{-1}) are treated as parameters. The weight on the reference deviation term QR(k) is denoted as a function of the time step as we can optionally define a terminal weight QR(N) similar to the P matrix in the LQR case. The rest of the problem formulation and variable definition remains the same as in Eq. (10.17).

Considering $N = M$ the problem transforms into Eq. (11.23):

$$\begin{aligned}
&\underset{u}{\text{Minimize}} \; \sum_{k=0}^{N} S_k^T \mathrm{QR}(k) S_k + \sum_{k=0}^{N-1} \Delta u_k^T R1 \Delta u_k \\
&\text{Subject to} \begin{cases} x_{k+1} = Ax_k + Bu_k \\ y_k = Cx_k \end{cases} && \forall k \in [0, N-1] \\
&\Delta u_k = u_k - u_{k-1}, && \forall k \in [0, N-1] \\
&S_k = y_i^{ref} - y_{i,k}, && \forall k \in [0, N] \\
&x_{\min} \leq x_k \leq x_{\max}, && \forall k \in [0, N] \\
&y_{\min} \leq y_k \leq y_{\max}, && \forall k \in [0, N] \\
&u_{\min} \leq u_k \leq u_{\max}, && \forall k \in [0, N-1] \\
&A_T x_N \leq b_T.
\end{aligned} \tag{11.23}$$

This problem can be approached in two different ways. The first is using Eq. (10.20) and reaching an mp-QP formulation similar to Eq. (10.24) where the parametric vector is defined as $\theta = [x_0^T, u_{-1}^T, y_i^{refT}]^T$ (Eq. (11.24)).

$$\begin{aligned} &\underset{u}{\text{Minimize}} && u^T H u + u^T Z\theta \\ &\text{Subject to} && Gu \leq W + S\theta \\ & && \text{CR}_A \theta \leq \text{CR}_b. \end{aligned} \tag{11.24}$$

Here we are more interested in the second approach that involves the reformulation of the state-space model and the definition of the problem as a standard LQR problem, discussed in the next session.

11.3.1 Reference Tracking to LQR Reformulation

Let the first-order derivative of the optimal action of a linear discrete time state-space model of the form $\{x_{k+1} = Ax_k + Bu_k, y_k = Cx_k\}$ with respect to time t be constant and equal to a variable $g(t)$. By assuming that the within a time-step of T_s length of the discrete time system, the optimal action does not change (piecewise constant) Eq. (11.25) can be defined.

$$\begin{aligned} &\frac{du(t)}{dt} = g(t), \\ &\int_{t_{k-1}}^{t_k} \frac{du(t)}{dt} dt = \int_{t_{k-1}}^{t_k} g(t)dt, \\ &\int_{u(t_{k-1})}^{u(t_k)} du = \int_{t_{k-1}}^{t_k} dt\ g_k, \\ &u(t_k) - u(t_{k-1}) = (t_k - t_{k-1})g_k, \\ &u_k - u_{k-1} = T_s g_k, \\ &\text{or equivalently} \\ &\Delta u_k = T_s g_k. \end{aligned} \tag{11.25}$$

Therefore, the state-space model in discrete time format can be posed as (Eq. (11.26)):

$$\begin{aligned} &x_{k+1} = Ax_k + Bu_{k-1} + T_s B g_k, \\ &y_k = Cx_k, \\ &u_k = u_{k-1} + T_s g_k, \\ &\text{or equivalently} \\ &\begin{bmatrix} x_{k+1} \\ u_k \end{bmatrix} = \begin{bmatrix} A & B \\ 0 & I \end{bmatrix} \begin{bmatrix} x_k \\ u_{k-1} \end{bmatrix} + \begin{bmatrix} T_s B \\ T_s \end{bmatrix} g_k, \\ &y_k = \begin{bmatrix} C & 0 \end{bmatrix} \begin{bmatrix} x_k \\ u_{k-1} \end{bmatrix}. \end{aligned} \tag{11.26}$$

Equivalently we may define $S_k = y_i^{ref} - y_{i,k} = y_i^{ref} - C_i x_k$ and $S_{k+1} = y_i^{ref} - y_{i,k+1} = y_i^{ref} - C_i A x_k - C_i B u_k$, where C_i denotes the rows of C with index i, i.e. the indices of the outputs that reference tracking is applied to. Given the definitions earlier, Eq. (11.27) can be defined.

$$
\begin{aligned}
&x_{k+1} = Ax_k + Bu_{k-1} + T_s B g_k,\\
&y_k = Cx_k,\\
&u_k = u_{k-1} + T_s g_k,\\
&S_{k+1} = S_k C_i(I - A)x_k - C_i B u_{k-1} - T_s C_i B g_k,\\
&y_k = Cx_k,\\
&\text{or equivalently}\\
&\begin{bmatrix} S_{k+1} \\ x_{k+1} \\ u_k \end{bmatrix} = \begin{bmatrix} I & C_i(I - A) & -C_i B \\ 0 & A & B \\ 0 & 0 & I \end{bmatrix} \begin{bmatrix} S_k \\ x_k \\ u_{k-1} \end{bmatrix} + \begin{bmatrix} -TsC_i B \\ T_s B \\ T_s \end{bmatrix} g_k,\\
&\qquad y_k = \begin{bmatrix} 0 & C & 0 \end{bmatrix} \begin{bmatrix} S_k \\ x_k \\ u_{k-1} \end{bmatrix}.
\end{aligned}
\tag{11.27}
$$

Note that:

- The definition of the output y_k is still necessary and relevant as it will play a role in the definition of (i) box constraints on the reference values and therefore on S_k and (ii) path constraints.
- The variable S_k is treated as a parameter (instead of y_i^{ref}) but we will show how this can be alleviated.

The reference tracking problem of Eq. (11.23), given the state-space re-formulation of Eq. (11.27), can be posed as an LQR of the form of Eq. (10.18) where the state vector is now defined as $\tilde{x}_k = \left[S_k^T,\ x_k^T,\ u_{k-1}^T\right]$. The overall problem formulation and matrix definition is presented in Eq. (11.28).

$$
\begin{aligned}
&\underset{u}{\text{Minimize}}\ \tilde{x}_N^T P \tilde{x}_N + \sum_{k=0}^{N-1} \tilde{x}_k^T Q \tilde{x}_k + \sum_{k=0}^{N-1} g_k^T R g_k\\
&\text{Subject to}\ \tilde{x}_{k+1} = \tilde{A}\tilde{x}_k + \tilde{B} g_k\\
&\qquad\qquad y_k = \tilde{C}\tilde{x}_k \qquad\qquad \forall k \in [0, N-1]
\end{aligned}
$$

$$\begin{aligned}
&\begin{bmatrix} S_{\min} \\ x_{\min} \\ u_{\min} \end{bmatrix} \le \tilde{x}_k \le \begin{bmatrix} S_{\max} \\ x_{\max} \\ u_{\max} \end{bmatrix}, && \forall k \in [0, N] \\
&y_{\min} \le y_k \le y_{\max}, && \forall k \in [0, N] \\
&g_{\min} \le g_k \le g_{\max}, && \forall k \in [0, N-1] \\
&A_T \tilde{x}_N \le b_T,
\end{aligned} \tag{11.28}$$

where

$$\begin{aligned}
\tilde{A} &= \begin{bmatrix} I & C_i(I-A) & -C_i B \\ 0 & A & B \\ 0 & 0 & I \end{bmatrix}, \\
\tilde{B} &= \begin{bmatrix} -TsC_i B \\ T_s B \\ T_s \end{bmatrix}, \\
\tilde{C} &= \begin{bmatrix} 0 & C & 0 \end{bmatrix}, \\
Q &= Q(k) = \begin{bmatrix} QR(k) & 0 & \cdots & 0 \\ 0 & 0 & \cdots & 0 \\ \vdots & \vdots & \ddots & \vdots \\ 0 & 0 & \cdots & 0 \end{bmatrix}, \quad \forall k \in [0, N-1], \\
R &= T_s^2 R1, \quad \text{given the definition of } g_k \text{ in Eq. (11.26)}, \\
P &= QR(N)\text{: the solution to the discrete time, Riccati equation} \\
S_{min} &= y_{i,\min} - y_{i,\max}, \\
S_{max} &= y_{i,\max} - y_{i,\min}, \\
g_{min} &= \frac{u_{\min} - u_{\max}}{T_s}, \\
g_{max} &= \frac{u_{\max} - u_{\min}}{T_s}.
\end{aligned} \tag{11.29}$$

The optimization problem described in Eqs. (11.28) and (11.29) corresponds to an LQR of the form of Eq. (10.18) while performing the operations of the problem in Eq. (11.23). Note that in the formulation of Eq. (11.29), the bounds on g_k can be chosen such that they restrict the change between consecutive control actions in the form $T_s \Delta u_{min} \le g_k \le T_s \Delta u_{max}$. It was shown in Chapter 10 that this can be reformulated to an mp-QP where the vector of the initial values of the states is treated as unknown but bounded parameters. In this case, the corresponding mp-QP takes the form of Eq. (11.30).

$$\begin{aligned} &\underset{u}{\text{Minimize}}\ u^T H u + u^T Z^* \theta^* \\ &\text{Subject to}\ Gu \leq W + S^* \theta^* \\ &\qquad\qquad \text{CR}_A^* \theta^* \leq \text{CR}_b. \end{aligned} \tag{11.30}$$

On the contrary to Eq. (11.24), here, θ^* corresponds to $\theta^* = [S_0^T, x_0^T, u_{-1}^T]^T$. It is though obvious that the first element of the parametric vector is a linear combination of independent parameters. Via linear manipulations this can be alleviated. In order to continue we will assume that the problem at hand corresponds to the reference tracking of j outputs of a system of m states and n input variables, therefore the overall parametric vector has $j + m + n$ elements. In Eq. (11.31) the subscript j on matrix M corresponds to a submatrix $M(:, 1 : j)$, the subscript m corresponds to a submatrix $M(:, j + 1 : j + m)$ and the subscript n corresponds to a submatrix $M(:, 1 + j + m : j + m + n)$. Therefore for any matrix in Eq. (11.30) where $M\theta^*$ is present, by substituting $S_0 = y_i^{ref} - C_i x_0$, the following reformulation can be derived, without loss of generality (Eq. (11.31)):

$$\begin{aligned} M\theta^* &= \begin{bmatrix} M_j & M_m & M_n \end{bmatrix} \begin{bmatrix} S_0 \\ x_0 \\ u_{-1} \end{bmatrix}, \\ M\theta^* &= \begin{bmatrix} M_j & M_m & M_n \end{bmatrix} \begin{bmatrix} y_i^{ref} - C_i x_0 \\ x_0 \\ u_{-1} \end{bmatrix}, \\ M\theta^* &= \begin{bmatrix} M_m - C_i M_j & M_n & M_j \end{bmatrix} \theta, \end{aligned} \tag{11.31}$$

where $\theta = \left[y_i^{ref,T}, x_0^T, u_{-1}^T \right]^T$, equivalently to Eq. (11.24). Therefore, Eq. (11.30) can be written in the exact form of Eq. (11.24), with $Z = \left[Z_m^* - C_i Z_j^* \; Z_n^* \; Z_j^* \right]$, $S = \left[S_m^* - C_i S_j^* \; S_n^* \; S_j^* \right]$ and $\text{CR}_A = \left[\text{CR}_{A,m}^* - C_i \text{CR}_{A,j}^* \; \text{CR}_{A,n}^* \; \text{CR}_{A,j}^* \right]$. Applying the techniques presented in Chapter 4 the explicit solution to the reference tracking problem of Eq. (11.23) can be derived.

11.3.2 Explicit Reference Tracking – An Example of mp-MPC

Consider the state space model described by the matrices of Eq. (11.32). Also consider that the first output is tracked to a set-point value y_1^{ref}. The bounds for states, inputs, and outputs is given in Eq. (11.33), assuming $N = M = 10$. Note that the bounds on the second output are path constraints:

$$A = \begin{bmatrix} 1 & 1 \\ 0 & 1 \end{bmatrix}, \ B = \begin{bmatrix} 0 \\ 1 \end{bmatrix}, \ C = \begin{bmatrix} 2 & 1 \\ 1 & 2 \end{bmatrix}, \tag{11.32}$$

$$\begin{aligned}
&\text{Bounds on states:} && \begin{bmatrix} -10 \\ -10 \end{bmatrix} \leq x_k \leq \begin{bmatrix} 10 \\ 10 \end{bmatrix}, && \forall k \in [0, 10], \\
&\text{Bounds on inputs:} && -1 \leq u_k \leq 1, && \forall k \in [0, 9], \\
&\text{Bounds on outputs:} && \begin{bmatrix} -12.5 \\ -3 \end{bmatrix} \leq y_k \leq \begin{bmatrix} 12.5 \\ 3 \end{bmatrix}, && \forall k \in [0, 10].
\end{aligned} \tag{11.33}$$

Following Eqs. (11.28) and (11.29), and assuming $T_s = 1$, the reformulated state space model is brought in the form of Eq. (11.34):

$$\begin{aligned}
\tilde{A} &= \begin{bmatrix} 1 & 0 & -2 & -1 \\ 0 & 1 & 1 & 0 \\ 0 & 0 & 1 & 1 \\ 0 & 0 & 0 & 1 \end{bmatrix}, \\
\tilde{B} &= \begin{bmatrix} -1 \\ 0 \\ 1 \\ 1 \end{bmatrix}, \\
\tilde{C} &= \begin{bmatrix} 0 & 2 & 1 & 0 \\ 0 & 1 & 2 & 0 \end{bmatrix}, \\
Q &= Q(k) = \begin{bmatrix} 1 & 0 & 0 & 0 \\ 0 & 0 & 0 & 0 \\ 0 & 0 & 0 & 0 \\ 0 & 0 & 0 & 0 \end{bmatrix}, \quad \forall k \in [0, 9], \\
R &= 0.01, \\
P &= \text{QR}(10) \\
&: \text{the solution to the discrete time Riccati equation}, \\
S_{min} &= -25, \\
S_{max} &= 25, \\
g_{min} &= -2, \\
g_{max} &= 2.
\end{aligned} \tag{11.34}$$

Following the procedure described be Eqs. (10.19)–(10.22), we derive the mp-QP formulation corresponding to the mp-MPC in the form of Eq. (10.24). Equations (B.2)–(B.6) in Section B.1 define the corresponding mp-QP matrices. Note that this is the minimal problem representation, without the quadratic term of the parameters on the objective function and after the removal of redundant constraints (if any).[1]

Having brought the problem into the standard mp-QP format and by applying the procedures and algorithms in Chapters 3 and 4, we can acquire the explicit solution for the constrained LQR problem.

11.3.3 Results and Validation

The solution to the mp-QP of Eq. (11.30) yields a solution of 527 critical regions using the connected graph algorithm. Every critical region has a distinct parametric optimal function set for the 10 optimization variables u. In Table 11.3 we present 2D projections of the critical regions graphically and the analytical expressions for the unconstrained critical region. Note that the definition of the critical region where the input is optimal is denoted by the constraints of the form $\text{CR} = \text{CR}_A(i)\theta \leq \text{CR}\ (i)_b$, the corresponding optimal input is given in the form of $u(\theta) = A(i)\theta + b(i)$ and the objective function value is given in the form $z(\theta) = \theta^T Q(i)\theta + c_i^T\theta + d_i$, where i is the critical region number and $\theta = \left[y_i^{ref,T}, x_0^T,\ u_{-1}^T\right]^T$ (Figures 11.5 and 11.6).

11.4 Moving Horizon Estimation

11.4.1 Multi-parametric Moving Horizon Estimation

Moving horizon estimation is an estimation method based on optimization that considers a limited amount of past data. One of the

1 Usually, mp-MPC problems suffer from a large number of redundant constraints. The removal of the constraints a priori can help with the solution procedure significantly.

Table 11.3 The partial analytical solution of the mp-MPC problem.

$$\mathrm{CR}_1 \begin{cases} CR = \begin{bmatrix} 0.23 & -0.46 & -0.86 & 0.01 \\ -0.23 & 0.46 & 0.86 & -0.01 \\ 0.61 & -0.62 & -0.50 & 0.03 \\ -0.61 & 0.62 & 0.50 & -0.03 \\ -0.37 & 0.74 & 0.56 & -0.04 \\ 0.37 & -0.74 & -0.56 & 0.04 \\ 0.81 & -0.31 & -0.49 & -0.03 \\ -0.81 & 0.31 & 0.49 & 0.03 \\ 0.89 & 0.34 & 0.29 & -0.02 \\ -0.89 & -0.34 & -0.29 & 0.02 \\ 1.00 & -0.05 & -0.05 & 0.00 \\ -1.00 & 0.05 & 0.05 & 0.00 \\ 0.00 & 0.00 & 0.00 & 1.00 \\ 0.00 & 0.00 & 0.00 & -1.00 \\ 0.00 & 0.45 & 0.89 & 0.00 \\ 0.00 & -0.45 & -0.89 & 0.00 \end{bmatrix} \begin{bmatrix} y^{ref} \\ x_0^1 \\ x_0^2 \\ u_{-1} \end{bmatrix} \leq \begin{bmatrix} 0.45 \\ 0.45 \\ 1.78 \\ 1.78 \\ 0.81 \\ 0.81 \\ 3.95 \\ 3.95 \\ 6.39 \\ 6.39 \\ 5.82 \\ 5.82 \\ 1.00 \\ 1.00 \\ 1.34 \\ 1.34 \end{bmatrix} \quad u(\theta) = \begin{bmatrix} 0.51 & -1.02 & -1.92 & -0.97 \\ -0.96 & 1.92 & 2.61 & -0.07 \\ 0.33 & -0.65 & -0.32 & 0.07 \\ 0.20 & -0.39 & -0.48 & -0.01 \\ -0.05 & 0.11 & 0.06 & -0.01 \\ -0.03 & 0.05 & 0.07 & 0.00 \\ 0.01 & -0.02 & -0.01 & 0.00 \\ 0.00 & -0.01 & -0.01 & 0.00 \\ 0.00 & 0.00 & 0.00 & 0.00 \\ 0.00 & 0.00 & 0.00 & 0.00 \end{bmatrix} \begin{bmatrix} y^{ref} \\ x_0^1 \\ x_0^2 \\ u_{-1} \end{bmatrix} \\ z(\theta) = \begin{bmatrix} y^{ref} \\ x_0^1 \\ x_0^2 \\ u_{-1} \end{bmatrix}^T \begin{bmatrix} 22572.30 & -4.77 & -0.91 & -1.05 \\ -4.77 & 9.53 & 1.83 & 2.10 \\ -0.91 & 1.83 & 4.28 & -0.81 \\ -1.05 & 2.10 & -0.81 & 1.10 \end{bmatrix} \begin{bmatrix} y^{ref} \\ x_0^1 \\ x_0^2 \\ u_{-1} \end{bmatrix} \end{cases}$$

The term CR corresponds to the critical region definition, the term k to the active set, the term $u(\theta)$ to the optimal parametric input function and the term $z(\theta)$ to the parametric objective function. Note that (i) the term Lagrange multipliers ($\lambda(\theta)$), (ii) the active set, and (iii) the fixed term of the optimal control action are omitted since this critical region corresponds to the unconstrained solution of the problem.

$\theta = \left[y_i^{ref,T}, x_0^T, u_{-1}^T\right]^T$

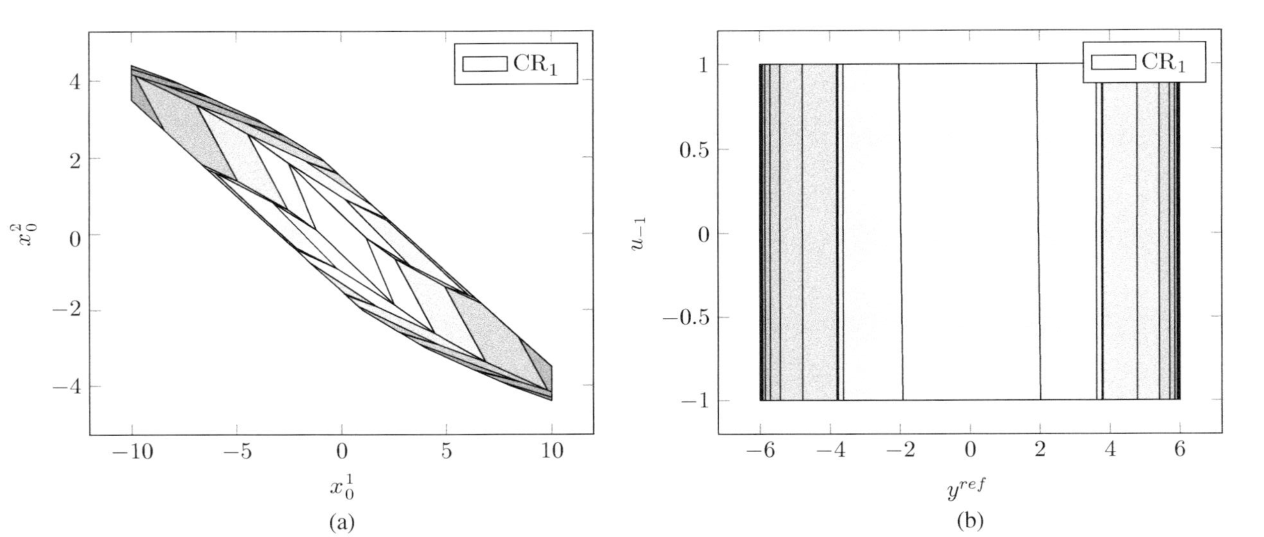

Figure 11.5 Indicative snapshots of the solution of the explicit reference tracking mp-MPC example problem. Note that the complete solution in a four-dimensional parameter space consists of 527 critical regions. The first critical region corresponds to the unconstrained control action law. (a) The projection of the parameter space on the initial state parameters. It consists of 47 critical regions is shown, each of which is associated with a different set of active constraints k_i. The legend shows only 1 out of the 47 critical regions to simplify the figure. (b) The projection of the parameter space on the previous optimal action and reference output. It consists of 25 critical regions is shown, each of which is associated with a different set of active constraints k_i. The legend shows only 1 out of the 25 critical regions to simplify the figure.

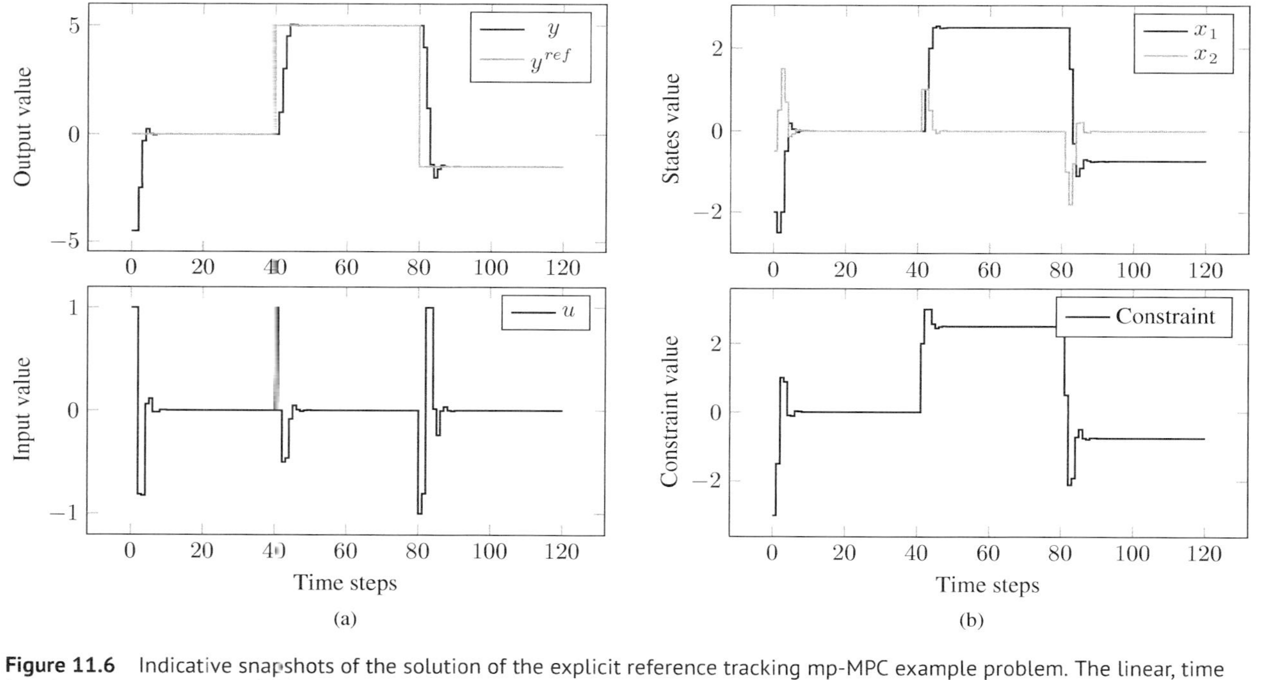

Figure 11.6 Indicative snapshots of the solution of the explicit reference tracking mp-MPC example problem. The linear, time invariant, state-space, model is simulated in a closed loop fashion. In this particular snapshot, the initial state values have been chosen to be $x_0^1 = -2$ and $x_0^2 = -0.5$. (a) Observe how the controller is able to stabilize the system around the set-points (0, 5 and −1.5). (b) Observe how the controller is able to stay within the constraint specification $| \begin{bmatrix} 1 & 2 \end{bmatrix} x_0^i| \leq 3, \forall i \in \{0, \dots, N-1\}$.

main advantages of moving horizon estimation is the possibility to incorporate system knowledge as constraints in the estimation. In MHE the system states and disturbances are derived by solving the following optimization problem [3–5]:

$$\begin{array}{ll} \min\limits_{\hat{x}_{T-N/T}, \hat{W}_T} & ||\hat{x}_{T-N/T} - \underline{x}_{T-N/T}||^2_{P^{-1}_{T-N/T-1}} - \\ & ||Y^{T-1}_{T-N} - O\hat{x}_{T-N/T} - \bar{c}bU^{T-2}_{T-N}||^2_{P-1} \\ & + \sum\limits_{k=T-N}^{T-1} ||\hat{w}_k||_{Q_k^{-1}} + ||\hat{v}_k||_{R_k^{-1}} \\ \text{s.t.} & \hat{x}_{k+1} = A\hat{x}_k + Bu_k + G\hat{w}_k \\ & y_k = C\hat{x}_k + \hat{v}_k \\ & \hat{x}_k \in X, \quad \hat{w}_k \in \Theta, \quad \hat{v}_k \in V, \end{array} \tag{11.35}$$

where T is the current time, $Q_k \succ 0$, $R_k \succ 0$, $P_{T-N/T-1} \succ 0$ are the covariances of w_k, v_k, x_{T-N} assumed to be symmetric, N is the horizon length of the MHE, Y^{T-1}_{T-N} is a vector containing the past $N+1$ measurements, and U^{T-1}_{T-N} is a vector containing the past N inputs. x, v, and w denote the variables of the system and $\hat{x}$, $\hat{v}$, and $\hat{w}$ denote the estimated variables of the system and $\hat{x}_{T/T-N}$ and $\hat{W}_T = W^{T-1}_{T-N}$ denote the decision variable of the optimization problem, the estimated state variable, and the noise, respectively. In multi-parametric moving horizon estimation (mp-MHE), the matrices Q_k, R_k, and $P_{T-N/T-1}$ are constant. In particular $P_{T-N/T-1} = P_{ss}$, which corresponds to the steady-state Kalman covariance matrix.

To obtain an mp-MHE formulation, the problem in Eq. (11.35) is reformulated as a multi-parametric programming problem:

$$\begin{array}{ll} \min\limits_{\hat{x}_{T-N/T}, \hat{W}^{T-1}_{T-N/T}} & \frac{1}{2}[\hat{x}_{T-N/T}T, \hat{W}^{T-1}_{T-N/T}]H \begin{bmatrix} \hat{x}_{T-N/T} \\ \hat{W}^{T-1}_{T-N/T} \end{bmatrix} \\ +\theta \cdot f \cdot \begin{bmatrix} \hat{x}_{T-N/T} \hat{W}^{T-1}_{T-N/T} \end{bmatrix} & \\ \text{s.t.} & K \cdot \begin{bmatrix} \hat{x}_{T-N/T} \\ \hat{W}^{T-1}_{T-N/T} \end{bmatrix} \leq k. \end{array} \tag{11.36}$$

The parameters of the multi-parametric programming problem (11.36) are the past and current measurements and inputs and the initial guess for the estimated states.

11.4.1.1 Current State

Unconstrained MHE: There are a few necessary steps that lead to incorporating the constrained MHE into robust MPC. The estimation error at the beginning of the horizon and at the current time have to be derived and the bounding set of the estimation error has to be obtained. Since the unconstrained moving horizon estimator is equivalent to the Kalman filter [3, 6], the estimation error and the bounding sets they generate should be equivalent so the Kalman filter can be used for comparison.

Constrained MHE: In order to formulate and solve the constrained moving horizon estimator with multi-parametric programming problem, the optimization problem is reformulated into the standard multi-parametric quadratic form. Previous work has been performed on reformulating the MHE with the filtered arrival cost [7] and with the smoothed update of the arrival cost [8].

11.4.1.2 Recent Developments

MHE with smoothed arrival cost: The formulation of the MHE with the smoothed arrival cost is still an open issue in literature. The optimization problem is reformulated into the standard multi-parametric quadratic form. The smoothed update of the arrival cost involves less parameters in the multi-parametric formulation of the MHE and hence it requires less computational effort to solve the mp-MHE than the equivalent estimation problem that utilizes the filtered arrival cost [9].

Simultaneous mp-MHE and mp-MPC: The implementation of explicit/multi-parametric MPC, and in general, MPC, is based on the assumption that the state values are readily available from the system measurements and also that the measurable output is free of noise influence. However, in reality, the measured output may be noisy and the system measurements do not offer this information directly – instead the state information needs to be inferred from the available output measurements by using a state estimator that obtains an estimate $\hat{x}$ of the real state x. The framework uses the constrained MHE that gives improved estimation results compared with the unconstrained estimators. The estimation error remains inside the calculated error set and hence the MPC guarantees to satisfy the constraints [10, 11].

Simultaneous mp-MHE and mp-MPC for biomedical applications: Biomedical systems are complex systems with a high degree of nonlinearity. Estimation techniques play an important role in such processes since some of the parameters and the states of the systems cannot always be measured directly from the system outputs. In most of the biomedical applications, the optimal policies rely on patient-dependent data that might be unavailable or computationally impossible to retrieve in a reasonable time frame. This makes simultaneous mp-MHE and mp-MPC an important ongoing research area that can deal with some critical issues especially on topics such as intravenous and volatile anesthesia, type-1 diabetes, and leukemia.

11.4.1.3 Future Outlook

Simultaneous mp-MHE and mp-MPC for periodic systems: Simultaneous mp-MHE and mp-MPC are areas that have been receiving more attention in the past years. The research work in this field has been acknowledged in several publications. One area that represents an important research is the design of simultaneous control and estimation schemes for periodic systems. The periodic nature and the presence of multiple control objectives make the control of such processes a challenging task.

mp-MHE for hybrid systems: The control of hybrid systems[2] represents a demanding challenge on its own. So far the multi-parametric moving horizon estimation has not been addressed in the context of hybrid systems. Current research will be focusing on multi-parametric MIQP algorithm and a step-by-step procedure for the derivation of offline multi-parametric hybrid controllers [12–14] and an integrated software (PAROC – PARametric Optimization and Control) for the general design, operational optimization, and control of process systems.

---

2 In this context the term hybrid systems is used to denote systems that involve both continuous and binary decision variables.

11.5 Other Developments in Explicit MPC

The material outlined in this chapter covers the standard use of explicit MPC for continuous and mixed-integer systems. However, the area of explicit MPC has featured several other developments, some of the most important ones are described in the following:

Continuous-time explicit MPC: The optimal control strategies for discrete-type systems of type (10.15) are determined off-line by solving multi-parametric programming problems of type (10.24). On the other hand, for systems with continuous-time dynamics, it is necessary to consider so-called multi-parametric dynamic optimization (mp-DO) problems, which lead to an infinite dimensional problem. Within the literature, two different strategies have been proposed to solve an mp-DO. One way is to transform the mp-DO problem into a finite-dimensional multi-parametric problems via control vector parameterization [15], while the other way is to solve the mp-DO problem directly in the infinite-dimensional form using variational approaches. While the theory presented earlier in this chapter is applicable to the finite-dimensional reformulation, for the infinite dimensional problems, it has been proposed to transform the optimization problems into a boundary value problem derived from the corresponding optimality conditions [16, 17].
These insights have led to the recent development of a unified framework, which combines the formulation of the control problem as an mp-DO with the tracking of the necessary conditions for optimality (NCO), namely, multi-parametric NCO-tracking [17]. The aim of this method is to convert an online dynamic optimization problem into a measurement based feedback control problem. This combination of mp-DO and NCO-tracking enables the relaxation of the fixed switching structure from an NCO-tracking perspective, as it constructs the critical regions which correspond to different optimal switching structures. This leads to a great reduction in the number of critical regions and the explicit solution of the continuous-time MPC problem.

Decentralized explicit MPC: The application of explicit MPC is often limited to the size of the problems under consideration. While in some cases it is the single system that requires a prohibitively large number of states or control variables, there are other cases where the system consists of several interconnected elements. The advantage of explicit MPC is that these elements can be solved independently, and then linked to each other by expressing the input of one element as the output of another. This has gained some recent interest in the research community, where the use of vertical and horizontal decentralization enables the use of explicit MPC, with its inherent advantages, for large and complex systems [18, 19]. This strategy was, for example, successfully applied in [20] for periodic systems and in [21] for combined heat and power systems.

References

1 Floudas, C.A. (1995) *Nonlinear and mixed-integer optimization: fundamentals and applications*, *Topics in chemical engineering*, Oxford University Press, New York.

2 Oberdieck, R. and Pistikopoulos, E.N. (2015) Explicit hybrid model-predictive control: the exact solution. *Automatica*, 58, 152–159, doi: 10.1016/j.automatica.2015.05.021. URL http://www.sciencedirect.com/science/article/pii/S0005109815002277.

3 Mayne, D.Q., Rawlings, J.B., Rao, C.V., and Scokaert, P. (2000) Constrained model predictive control: stability and optimality. *Automatica*, 36 (6), 789–814, doi: 10.1016/S0005-1098(99)00214-9. URL http://www.sciencedirect.com/science/article/pii/S0005109899002149.

4 Rawlings, J.B., Mayne, D.Q., and Diehl, M.M. (2017) *Model predictive control: theory and design*, Nob Hill Publishing, Madison, WI, 2nd edn.

5 Tenny, M. (2002) *Computational strategies for nonlinear model predictive control*, Ph.D. thesis, University of Wisconsin-Madison, Wisconsin-Madison.

6 Findeisen, P.K. (1997) *Moving horizon state estimation of discrete time systems*, Ph.D. thesis, University of Wisconsin-Madison, Wisconsin-Madison.

7 Darby, M.L. and Nikolaou, M. (2007) A parametric programming approach to moving-horizon state estimation. *Automatica*, 43 (5), 885–891, doi: 10.1016/j.automatica.2006.11.021. URL http://www.sciencedirect.com/science/article/pii/S0005109807000283.

8 Voelker, A., Kouramas, K., and Pistikopoulos, E.N. (2010) Unconstrained moving horizon estimation and simultaneous model predictive control by multi-parametric programming, in *UKACC International Conference on Control 2010*, pp. 1–6, doi: 10.1049/ic.2010.0443.

9 Voelker, A., Kouramas, K., and Pistikopoulos, E.N. (2010) Simultaneous constrained moving horizon state estimation and model predictive control by multi-parametric programming, in *2010 49th IEEE Conference on Decision and Control (CDC)*, pp. 5019–5024, doi: 10.1109/CDC.2010.5717762.

10 Lambert, R., Nascu, I., and Pistikopoulos, E.N. (2013) Simultaneous reduced order multi-parametric moving horizon estimation and model predictive control, in *Dynamics and Control of Process Systems*, Elsevier, IFAC, IFAC proceedings volumes, pp. 45–50, doi: 10.3182/20131218-3-IN-2045.00071.

11 Voelker, A., Kouramas, K., and Pistikopoulos, E.N. (2010) Simultaneous state estimation and model predictive control by multi-parametric programming. *Computer Aided Chemical Engineering*, 28 (C), 607–612.

12 Dua, V., Bozinis, N.A., and Pistikopoulos, E.N. (2002) A multi-parametric programming approach for mixed-integer quadratic engineering problems. *Computers and Chemical Engineering*, 26 (4–5), 715–733, doi: 10.1016/S0098-1354(01)00797-9. URL http://www.sciencedirect.com/science/article/pii/S0098135401007979.

13 Pistikopoulos, E.N. (2009) Perspectives in multiparametric programming and explicit model predictive control. *AIChE Journal*, 55 (8), 1918–1925, doi: 10.1002/aic.11965. URL http://dx.doi.org/10.1002/aic.11965.

14 Nascu, I., Lambert, R.S.C., Krieger, A., and Pistikopoulos, E.N. (2014) Simultaneous multi-parametric model predictive control and state estimation with application to distillation

column and intravenous anaesthesia, in *24th European Symposium on Computer Aided Process Engineering, Computer Aided Chemical Engineering*, vol. 33 (eds J. Jaromír Klemeš, P.S. Varbanov, and P.Y. Liew), Elsevier, pp. 541–546, doi: 10.1016/B978-0-444-63456-6.50091-0.

15 Sakizlis, V. (2003) *Design of model based controllers via parametric programming*, Ph.D. thesis, Imperial College, London.

16 Sakizlis, V., Perkins, J.D., and Pistikopoulos, E.N. (2005) Explicit solutions to optimal control problems for constrained continuous-time linear systems. *IEEE Proceedings: Control Theory and Applications*, 152 (4), 443–452, doi: 10.1049/ip-cta:20059041.

17 Sun, M., Chachuat, B., and Pistikopoulos, E.N. (2016) Design of multi-parametric NCO tracking controllers for linear dynamic systems. *Computers and Chemical Engineering*, 92, 64–77, doi: 10.1016/j.compchemeng.2016.04.038. URL http://www.sciencedirect.com/science/article/pii/S009813541630134X.

18 Spudic, V., Jelavic, M., and Baotic, M. (2012) Explicit model predictive control for reduction of wind turbine structural loads, in *2012 IEEE 51st IEEE Conference on Decision and Control (CDC)*, pp. 1721–1726, doi: 10.1109/CDC.2012.6426490.

19 Spudic, V. and Baotic, M. (2013) Fast coordinated model predictive control of large-scale distributed systems with single coupling constraint, in *2013 European Control Conference (ECC)*, Zurich, Switzerland 17–19 July 2013, pp. 2783–2788.

20 Papathanasiou, M.M., Avraamidou, S., Steinebach, F., Oberdieck, R., Mueller-Spaeth, T., Morbidelli, M., Mantalaris, A., and Pistikopoulos, E.N. (2016) Advanced control strategies for the multicolumn countercurrent solvent gradient purification process (MCSGP). *AIChE Journal*, 62 (7), 2341–2357, doi: 10.1002/aic.15203. URL http://dx.doi.org/10.1002/aic.15203.

21 Diangelakis, N.A., Avraamidou, S., and Pistikopoulos, E.N. (2016) Decentralized multiparametric model predictive control for domestic combined heat and power systems. *Industrial and Engineering Chemistry Research*, 55 (12), 3313–3326, doi: 10.1021/acs.iecr.5b03335. URL http://dx.doi.org/10.1021/acs.iecr.5b03335.

12

PAROC: PARametric Optimization and Control

12.1 Introduction

The integration of detailed modeling, design and operational optimization, controller design, and scheduling planning policies are core process systems engineering challenges. While high-fidelity modeling and dynamic simulation have been becoming standard engineering tools, with software systems such as ASPEN Plus® [1] and gPROMS® [2] becoming standard platforms, the same is not true for the integration of design, control, and scheduling. It is interesting to note that despite major advances in the areas of design and control, scheduling and control, and model-based advanced control, there is currently (i) no generally accepted methodology and/or "protocol" for such an integration and (ii) not any commercially available software (or even in a prototype form) system to fully support such an activity. It is clear from Tables 12.1 and 12.2 that a plethora of attempts has been made to integrate design and/or operational optimization with advanced control schemes. Different control approaches and optimization methods and tools have contributed greatly towards this direction. Despite this fact, there has not been a coordinated approach to systematically tackle the design and control optimization problem, the operational and control optimization problem, or the unification of design.

Multi-parametric Optimization and Control, First Edition.
Efstratios N. Pistikopoulos, Nikolaos A. Diangelakis, and Richard Oberdieck.

Table 12.1 Interaction of design and control – indicative list.

Authors	Contributions
Lee *et al.* [3]	Introduction to design and control
Narraway *et al.* [4]	Steady state and dynamic economics
Luyben and Floudas [5]	Superstructure of design alternatives into MINLP
Mohideen *et al.* [6]	Economically optimal design and control
van Schijndel and Pistikopoulos [7]	Review on process design and operability
Sakizlis *et al.* [8]	Review on design and control
Flores-Tlacuahuac and Biegler [9]	MIDO transformed to MINLP
Würth *et al.* [10]	Dynamic optimization and NMPC
Yuan *et al.* [11]	Review on design and control
Diangelakis *et al.* [12]	Sequential design optimization and mp-MPC. Application on a residential cogeneration systems
Liu *et al.* [13]	Optimal design and operation of distributed systems with focus on the trade-offs between modeling accuracy and computational complexity and efficiency
Sakizlis *et al.* [14]	Simultaneous mp-MPC and online design optimization. Case studies on binary distillation column and evaporator

Here, we present a comprehensive framework that enables the representation and solution of demanding model-based operational optimization and control problems. It is based on the derivation of advanced model based multi-parametric optimization strategies using the techniques presented in the previous chapters. We show an integrated procedure featuring "high fidelity" modeling, approximation techniques, and optimization-based strategies, including multi-parametric programming. The key aspect of the framework is to provide a common basis for (i) the integration of design and control optimization, (ii) the integration of operational and control optimization, and (iii) the grand unification of design,

Table 12.2 Integration of scheduling and control – indicative list.

Authors	Contributions
Shobrys and White [15]	Interactions of planning scheduling and control and their impact to decision making in process industry operations
Mahadevan *et al.* [16]	Robust control for the targeting of transition times in scheduling of polymerization operation
Chatzidoukas *et al.* [17]	Impact of control structure on process operability, product quality optimization and time optimal grade transition
Chatzidoukas *et al.* [18]	Integration of production scheduling and optimal grade transition profiles with a MIDO approach
Nystrom *et al.* [19]	Production optimization through determination of transition trajectories, operating points and manufacturing sequence
Flores-Tlacuahuac and Grossmann [20, 21]	Simultaneous cyclic scheduling and control via reformulating an MIDO problem into an MINLP
Harjunkoski *et al.* [22]	Discussion on the problems arising from the integration of production scheduling and control and its implementation
Biegler and Zavala [23]	Real-time optimization and control for decision making via the formulation and efficient solution of NLP
Subramanian *et al.* [24]	Distributed MPC and cooperative MPC to integrate scheduling objectives with process operation constraints
Subramanian *et al.* [25]	MI scheduling problem formulation based on state-space models
Zhuge and Ierapetritou [26]	Continuous-time event-point formulation scheduling incorporating explicit constraints derived from mp-MPC to target complexity
Kopanos and Pistikopoulos [27]	Reactive scheduling of a state-space representation based system using multi-parametric programming
Baldea and Harjunkoski [28]	A comprehensive systematic review of the integration of scheduling and control
You and Grossmann [29]	Supply chain optimization under uncertainty via multi-period MINLP

operational, and control optimization in a consistent manner. A key advantage of the proposed framework is its ability to adapt to different classes of problems, in an effortless manner through a prototype software platform PAROC (PARametric Optimization and Control). The latter offers, among others, the great advantage of interoperability between advanced modeling software packages (PSE's gPROMS ModelBuilder) and MATLAB®-based tools for model approximation and multi-parametric model-based controller and state estimator design tools.

12.2 The PAROC Framework

In this section the PAROC framework is described in detail and depicted in Figure 12.1.

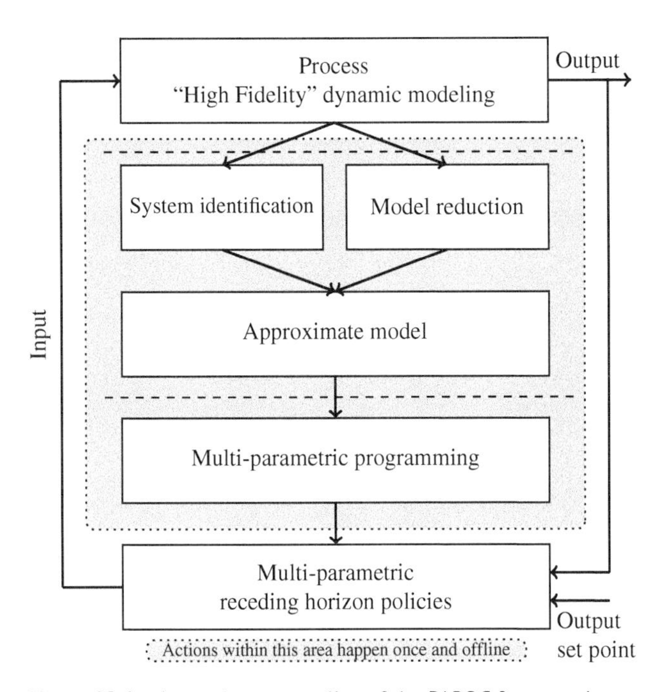

Figure 12.1 A step-by-step outline of the PAROC framework.

12.2.1 "High Fidelity" Modeling and Analysis

The first step of the PAROC framework is "high fidelity" modeling and analysis. In particular, the scope is to (i) develop a "high fidelity" model of the process [30, 31], (ii) analyze the original problem, e.g. using global sensitivity analysis [32–34], and (iii) perform parameter estimation and dynamic optimization of the developed model. Within our framework, the modeling software PSE's gPROMS ModelBuilder [2] is used, as it either provides the aforementioned tools directly or allows their implementation via gO:MATLAB, a connection tool between MATLAB and gPROMS.

12.2.2 Model Approximation

Although it is possible to use a "high fidelity" model for optimal design decisions, its complexity may usually render its direct use for the development of model-based strategies computationally expensive. Consequently, it may be necessary to simplify the representation of the model while compromising its accuracy. In PAROC this is addressed by the following two approaches:

System identification: A series of simulations of the "high fidelity" model for different initial states is used to construct a meaningful linear state-space model of the process using statistical methods. One of the most widely applied tools within this area is the System Identification Toolbox from MATLAB.

Model-reduction techniques: While system identification relies on the user in terms of interpretation of the data and processing of the results, model-reduction techniques somewhat "automate" the reduction process based on formal techniques (see Table 12.3 for recent contributions).

We devote Section 12.2.2.1 with users' perspective notes on the Model Approximation step as it is often the most deterministic factor of accuracy between the original process model and the resulting optimization strategies.

12.2.2.1 Model Approximation Algorithms: A User Perspective Within the PAROC Framework

Deriving a linear state-space model representation (in continuous or discrete time format) from a process "high-fidelity" model is

Table 12.3 Model approximation of multi-parametric model-predictive control – recent contributions.

References	Key contribution
Narciso and Pistikopoulos [44]	Combination of linear model reduction and linear multi-parametric model-predictive control (mp-MPC)
Rivotti *et al.* [45]	A model order reduction via empirical gramians [46] is combined with an mp-MPC algorithm
Lambert *et al.* [36]	Using Monte Carlo integrations, N step ahead affine representations are created

commonly referred to as model approximation. The procedure involves the reduction of complexity of the "high-fidelity" model while attempting to preserve its accurate representation of the process at hand. Clearly, a trade-off between the two is of the essence in this procedure. A large variety of the approximation method s exists in open literature involving the following:

- Model-based (piece-wise) linearization of the nonlinearities commonly present in the "high-fidelity" model around one or multiple predefined points [35], typically coupled with
- model order reduction techniques, in cases where the state vector size of an already linear (or linearized) model becomes a computational burden in the realm of multi-parametric programming [36–38].
- Data-driven methods where sets of input-output data are utilized to derive linear or nonlinear surrogate models, standalone or in tandem with equations from the "high-fidelity" model (black box and gray box models techniques are of the essence here [39, 40])
- commercially available algorithms that utilize one or various of the aforementioned characteristics (e.g. The System Identification Toolbox® of MATLAB®.).

Although interesting, analyzing the in-depth characteristics of each one of those categories would defeat the purpose of this chapter. Therefore, here we present the main underlying theoretical principles and basic usage guidelines from a PAROC user perspective for data-driven model approximation approaches. This is presented through the System Identification Toolbox® of MATLAB® and concerns mainly the "Subspace Algorithms for the Identification of Combined Deterministic – Stochastic Systems," commercially known as *N4SID*' [37]. N4SID is the most commonly used tool to derive state-space models within the PAROC framework, followed by "prediction error estimation methods" (PEM), also touched upon here.

12.2.2.1.1 Model Approximation: An Outline

In Figure 12.2 the model approximation procedure is presented and discussed in context with the PAROC framework. The procedure is iterative and repeats itself until a satisfactory model is found. The tasks described here take place within the System Identification Toolbox.

Data Collection The first user-performed step within the model approximation step of the PAROC framework is the collection of data from the computational experiment, in this case the "high-fidelity" model. The set of Input/Output (I/O) data is exchanged between gPROMS, which typically is the modeling software and MATLAB where the optimization and control problems are formulated and solved.[1] The form of the I/O data depends on the original model and the purpose of the control. For example, in order to build a controller that maintains an output set-point while rejecting disturbances, the manipulated input signal u is advisable to be consisting of step changes (including saturated actions) and the disturbance signal d is advisable to be in the form of white noise, i.e. the I/O data should mimic the ideal behavior of the controller. This is true for both data sets i and j in Figure 12.2.

Polish and Present Data The presentation of the I/O data set is the first task of the System Identification Toolbox. It is useful in cases

1 The dynamic degrees of freedom of the original model are treated as the Input set in this context. These can be manipulated inputs (which are ultimately optimized) or disturbances.

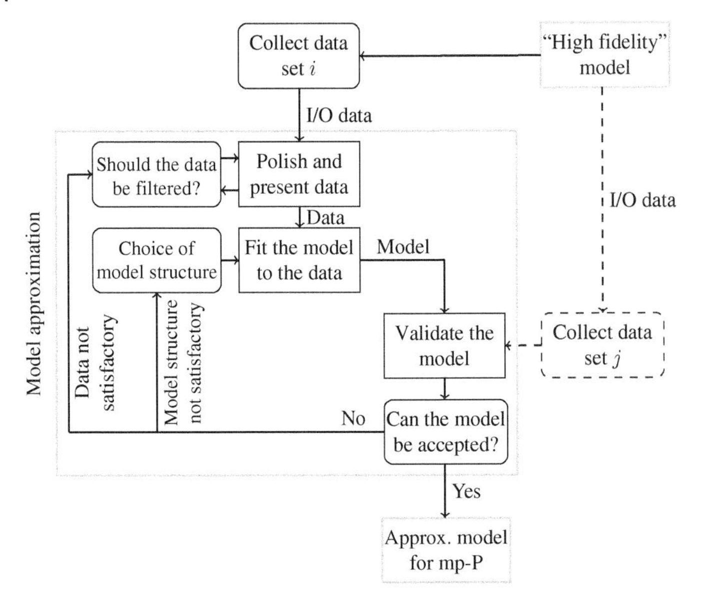

Figure 12.2 An outline of the model approximation procedure typically used within the PAROC framework. The gray boxes correspond to steps of the PAROC framework. The dashed lines and boxes correspond to optional actions. Rounded boxed correspond to actions performed by the user. Regular boxes correspond to actions performed by the computer. Source: Part of the figure has been adapted from Ljung [38].

where the data originates from an experimental procedure as its main task is the identification of outliers, which are automatically disregarded from the steps to follow. A graphical representation of the data set is provided where the user can inspect the behavior of the output data based on changes of the input signal. The user's task in this step is to verify that the data that is about to be used for the approximation is usable. The main concern here is the violation of physical properties of substances and chemical compounds, for example, data from a simulation that results into liquid water temperature higher than 100 degC in atmospheric pressure, should not be used. In that case, the input that causes that effect should be restrained to values that correspond to real operation or the original model should account for phase change.

Data Filtering Data filtering can refer to multiple procedures in this case. Typically, data scaling, data mean removal, data trend removal, and restraining the data set are actions that take place in this step. Restraining the data set is very helpful even during the first iteration of the procedure. It allows the user to choose the "data window" that describes the "high-fidelity" model at hand more accurately. Typically the user can disregard data time steps that correspond to specific events such as the start-up or the shut-down of the process and focus on the main operation (for continuous processes). In case of batch processes, this feature can be used to distinguish certain sequential process events that may require different approximate models in order to be approximated accurately. For example, a PVC polymerization reactor's operation significantly differs during startup and shutdown compared with its operation under the desirable polymerization temperature. In such cases, a set of I/O throughout the operating cycle can be "cut" into three distinct parts and handled accordingly. Such a restraining action is advisable in the first step of the approximation procedure. Data scaling is an important aspect of the model approximation procedure as well as the next steps in the PAROC framework. The vast majority of the available solvers and algorithmic numerical procedures guarantee better approximations when the models at hand (static or dynamic) are scaled within a certain range, which varies with respect to the solver. This includes the System Identification Toolbox gPROMS, CPLEX®as well as linprog and quadprog of MATLAB, the main tools within the PAROC framework. Therefore, for a flow of water that varies from 0 to 5 and results into a temperature change from 75 °C to 100 °C, the user is advised to scale the data into a uniform scale (i.e. both in the range of 0–1 or similar). Removing the data means and trends of the data are practices that should be considered after the initial data set fails to produce a reliable approximate model. Although removing the means may have a similar effect on the procedure as scaling the data, the trend removal is mainly effective on experimentally (not computationally) acquired data sets. Both actions require a reformulation of the final model to include for the filtering. The latter can have implications on the steps that follow within the PAROC framework as it can create extra terms in the approximate models that need to be accounted for.

Choosing the Model Structure In general, the model structure is a choice that needs to be made after taking a series of criteria into account. For example, the approximation of a nonlinear neural network can be in some cases very appealing and fitting to the data set at hand but the purpose of the approximation is defeated as the complexity reduction compared with the original model is very little, similarly for gray and black box approaches mentioned earlier [39, 40]. Given the application of receding horizon multi-parametric programming based techniques that follow the approximation step, the model structure choice becomes very limited. The two available choices are linear state-space models and ARX models of polynomial order 1 (linear ARX models). Although the latter has been used to demonstrate the applicability of multi-parametric receding horizon policies in [41], it is rarely used. The main reason is the availability of different techniques and approaches based on the form of the former. A number of decisions need to be made by the user for the derivation of a suitable state-space model representation. The ones discussed here include (i) the number of states, (ii) the discrete or continuous time representation, (iii) the inclusion of the output disturbance component K, (iv) the inclusion of feedthrough to the output, and (v) the focus of the model.

The Number of States This decision is the most fundamental decision with regards to the state-space model structure. The user needs to decide the number of linear states in which the approximate model estimation will take place. From a modeling point of view, the singular values of the state components are crucial. The singular value corresponding to order n is a measure of how much the nth component of the state vector contributes to the input–output behavior of the model. A reasonable choice of model order n is one where the singular values of components $n+i$ are small compared with those of components $n-j$, where i,j natural numbers and $j \leq n-1$. The System Identification Toolbox provides the ability to test multiple lengths of the state vector and determine their singular value. It also suggests the vector length for which the previous qualitative criterion is implicitly satisfied. Typically, the suggested state vector length tends to be high when approximating highly nonlinear systems. In the case where the user relies on the automated suggestion, it should be clear that a certain range of state vector lengths is tested at a time to reduce the computational effort, i.e. the Toolbox automatically tests vector sizes from 1 to 10. The user can change those bounds but a large range will result into significant computational times at best and failure to compute at worst. The user can test different ranges as they see fit.

Information from the original "high-fidelity" model can be very useful here. The number of states of the approximate model can be used as starting point for the state vector length determination procedure. It should be noted though that there is no guarantee that the approximate model state corresponds to the original model states. In this regard there exists a case where this can indeed be guaranteed. When the approximate and "high-fidelity" models are square in the state and output domain and the observation matrix is the identity matrix (an attribute which can be enforced *a posteriori* to the approximation procedure), then the approximate model states and the "high-fidelity" model states coincide. It is clear from the aforementioned that the usage of a state estimator during the online implementation of the receding horizon policies is often necessary.

In the context of PAROC, the user is urged to aim for the smallest possible number of states in the approximate model. This is necessary to ensure a reasonable number of constraints in the multi-parametric programming problem as well as a small number of parameters. The same is true for the online application of deterministic problems as well when the optimization procedure needs to be repeated within very little time.

Discrete or Continuous Time Representation A linear state-space model can be represented in both the discrete or the continuous time domain. The vast majority of algorithms and techniques for handling explicitly state-space models assume a discrete time model, although there have been significant advancements in the continuous time domain as well [42, 43]. Furthermore, the form of the I/O data is in the discrete time domain. It would be therefore intuitive to aim for the discrete time approximate state-space models in the discretization step suggested by the I/O data. Counter-intuitively, it is here suggested to aim for the continuous time state-space model representation, although ultimately a discrete model will be used for the receding horizon optimization policies. This is preferable because:

- The discretization procedure results always in a single model representation that depends on the discretization step while the opposite is not true. Based on the method (namely, zero-order hold on the inputs, linear interpolation of inputs, bilinear (Tustin) approximation, matched pole-zero method), the continuous time model can vary thus creating a reverse discretization mismatch error.
- Discretizing a continuous-time state-space model can affect the effectiveness of the receding horizon optimization policy. The discretization step can be re-adjusted prior to the formulation of the multi-parametric programming problems to help account for specific time lengths without changing the output horizon of the policy (i.e. an output horizon of 10 on a state-space model with discretization of 1 s covers a total of 10 s of operation. The same can be achieved for an output horizon of 8 and a discretization step of 1.25 s). This is advisable in cases where the process itself permits less tight control intervals and the size reduction of the multi-parametric programming problem is needed.

Ultimately, it is the choice of the user to acquire an approximate model in continuous or in discrete time format, based to the aforementioned criteria.

The Disturbance Component K The choice is fundamental in the approximation of linear state-space model when the mismatch between the "high fidelity" model and the approximate model is (expected to be) significant. The disturbance component K appears in the state space as shown in Eq. (12.1) (representation in discrete time format).

$$
\begin{aligned}
x_{T+1} &= Ax_T + Bu_T + Ky^{real}, \\
y_T &= Cx_T + Du_T + e, \\
e &= y^{real} - y_1.
\end{aligned}
\tag{12.1}
$$

The existence of the component K can drastically improve the fit of the approximate model to the I/O data, which is the main reason to be included in the state-space representation, but has a significant shortcoming. The online application of the receding horizon policy, especially in the case of the multi-parametric model predictive control (mp-MPC), requires the simultaneous usage of an advanced state estimator such as the moving horizon estimator (MHE) as shown in Chapter. In order for such a scheme to be part of the simultaneous approach in control, scheduling, and design optimization that would require not only the multi-parametric solution of the problem but also the design dependency of the multi-parametric moving horizon estimator (mp-MHE), it is this that leads to the exclusion of the component K throughout the thesis.

The Output Feedthrough This refers to the inclusion or not of the term Du_T at the output of the approximate state-space model of Eq. (12.1). The nature of the original "high-fidelity" model and especially the effect that the input has on the output is the main criterion for this choice. If, for example, an impulse of the u vector affects the output in an impulse way at the same discrete time step that the former happens, then there is high probability that the inputs algebraic effect on the output is greater than the differential effect (manifested through the term x_T). Furthermore, the terms $Cx_T + Du_T$ are very useful for imposing constraints between states and inputs via the output vector, a very common practice in MPC and rolling horizon optimization scheduling approaches.

The Focus of the Model For the purpose of this thesis and the rolling horizon optimization policies, the focus of the approximate model is simulation. This becomes important later on when validating the model against the I/O data is regarded.

12.2.2.1.2 Model Fitting

The model fitting is a task that is performed automatically by the System Identification Toolbox based on (i) the I/O data set i, (ii) the choices made by the user up to this point, and (iii) the algorithm of choice. In the context of the PAROC framework, the N4SID and the PEM algorithms are commonly used.

The N4SID algorithm is especially useful for identifying high-order multivariable systems for which it is not trivial to find a useful parameterization. The latter is of great importance in classical identification methods, such as PEM, which means that controllability and/or observability indices *a priori* knowledge is needed [37, 38]. Furthermore, the N4SID algorithm is non-iterative and does not include any nonlinear optimization step. The latter is of greater importance as they are not sensitive to initial conditions a characteristic that PEM algorithms suffer from, often resulting to non-convergence. Non-zero initial states can be handled by the N4SID algorithm the exact same way as zero states as they do not need an extra parameterization, a characteristic that the PEM algorithm inherently has.

The maximum likelihood method is the basis for the family of PEM algorithms, including the one discussed here [38]. The basic pros and cons are listed in the following:

- *Pro*: A wide spectrum of model structures can be used.
- *Pro*: The asymptotic properties of the derived models are guaranteed.
- *Con*; An explicit parameterization of the model is required.
- *Con*: The best output prediction fit relies on nonlinear optimization that can be computational expensive and results into local minima with severe effects on the model.

The fit of the model to the I/O data in both cases is defined by Eq. (12.2). The purpose of the approximation algorithms is to minimize the fit under the constraints imposed by the user choices.

$$\text{Fit} = \left[1 - \frac{||y^{real} - y^{sim}||^2}{||y^{real} - \bar{y}||^2}\right] 100, \tag{12.2}$$

where y is the measured output, y^{sim} is the simulated model output, and $\bar{y}$ is the mean output value. The length of the output vector over which the fit is defined can be changed by the user, and it can typically reflect the length of the output horizon to be used in the receding horizon policies later on.

12.2.2.1.3 Model Validation and Accepting the Model

The model validation has two parts. In the first part, the Toolbox automatically performs analyses on (i) the frequency and transient response of the approximate model, (ii) the noise spectrum, and (iii) zero pole cancellation. In section 17.2 of [38], an excellent analysis on the practical use of these features is provided within a general model approximation procedure (here omitted for brevity).

As a rule of thumb, a model that (i) yields a good fit (typically 80% or higher, but could vary with respect to the process at hand), (ii) is stable, (iii) does not have pole-zero cancellation (including the confidence intervals), (iv) its cross-correlation between inputs and residuals does not go significantly outside the confidence intervals, and (v) has less than a size 10 states vector[2] can be accepted for multi-parametric based receding horizon optimization policies.

The second part of the model validation includes an I/O data set j provided by the "high-fidelity" model. The j set of data is compared against the approximate model in open-loop to test the fit of the approximate model in a similar fashion as above. If the approximate model has a satisfactory performance with respect to both the automated criteria and fits a different set of I/O data well, then it can be accepted as a candidate for the formulation and solution of multi-parametric programming policies. Otherwise, the user needs to determine whether the structure of the approximate model of the I/O data set is problematic and repeat the procedure.

12.2.3 Multi-parametric Programming

After the model approximation step, a state space model is obtained, which is used for the development of receding horizon

2 The size of the states should not to be confused with the number of parameters present in the multi-parametric programming formulation. The number 10 usually corresponds to model sizes that have been computationally verified to be solvable within a reasonable amount of time on a regular computer via POP. The further development of the software as well as the usage of computer clusters and multi-core programming has shown significant progress to that direction (see Chapter 8).

Table 12.4 Different classes of multi-parametric programming problems encountered within PAROC.

Cost function System class	mp-P	Problem formulation	Chapters
1/∞-norm Continuous system	mp-LP	$z(\theta) = \min_x \; (H\theta + c)^T x$ s.t. $Ax \leq b + F\theta$, $x \in \mathbb{R}^n$	2 and 4
2-norm Continuous system	mp-QP	$z(\theta) = \min_x \; (Qx + H\theta + c)^T x$ s.t. $Ax \leq b + F\theta$, $x \in \mathbb{R}^n$	3 and 4
1/∞-norm Hybrid system	mp-MILP	$z(\theta) = \min_\omega \; (H\theta + c)^T \omega$ s.t. $Ax + Ey \leq b + F\theta$, $x \in \mathbb{R}^n,\ y \in \{0,1\}^p$, $\omega = [x^T\ y^T]^T$	5 and 7
2-Norm Hybrid system	mp-MIQP	$z(\theta) = \min_\omega \; (Q\omega + H\theta + c)^T \omega$ s.t. $Ax + Ey \leq b + F\theta$, $x \in \mathbb{R}^n,\ y \in \{0,1\}^p$, $\omega = [x^T\ y^T]^T$	6 and 7

Note that the parameters $\theta \in \Theta = \{\theta \in \mathbb{R}^q | \theta_l^{min} \leq \theta_l \leq \theta_l^{max}, l = 1, \ldots, q\}$ for all cases.

policies. The calculation of such policies, e.g. in the form of control laws or scheduling policies, traditionally requires the online solution of an optimization problem, which might be computationally infeasible [47]. Therefore, the PAROC framework employs multi-parametric programming, where the optimization problem is solved offline as a function of a set of parameters. In addition, depending on the cost function and the characteristic of the system considered, the complexity of the optimization problem changes considerably.

The theoretical developments and techniques presented in Chapters 2–7 are applied in this step for the different classes of optimization problems that may be encountered, a brief overview of which is given in Table 12.4.

12.2.4 Multi-parametric Moving Horizon Policies

While multi-parametric programming has been applied in a variety of areas, a key application lies in the offline calculation of moving horizon policies such as control laws and scheduling policies. The underlying idea is to consider the states of the system as parameters, and thus solve the optimization problem over a range of admissible states.

Remark 12.1 In addition, measured disturbances, if present, are also considered as parameters as well as state-space and model mismatch and the output set-point as shown in detail in Chapters 10 and 11.

The most common problems encountered in this step include multi-parametric model predictive controllers (mp-MPCs), mp-MHE, and multi-parametric Receding Horizon optimization policies (mp-RHO). The latter are treated as $1/\infty$-norm continuous or hybrid systems.

12.2.5 Software Implementation and Closed-Loop Validation

12.2.5.1 Multi-parametric Programming Software

In conjunction with the aforementioned theoretical developments, PAROC provides software solutions to key aspects of the framework (the interested reader can find software tools by visiting http://parametric.tamu.edu/PAROC/). In particular, it is seamlessly connected to POP for the formulation and solution of multi-parametric programming problems. Based on POP, it contains state-of-the-art algorithms that allow for an efficient solution of multi-parametric linear programming (mp-LP), multi-parametric quadratic programming (mp-QP), multi-parametric mixed-integer linear programming (mp-MILP), and multi-parametric mixed-integer quadratic programming (mp-MIQP) problems. Furthermore, its interconnection with gPROMS ModelBuilder (see Section 12.2.5.2) makes the use of the PAROC framework straightforward and allows for an intuitive approach for design, operation, and control problems. For the details of POP, see Chapter 8.

12.2.5.2 Integration of PAROC in gPROMS® ModelBuilder

The developed multi-parametric moving horizon policies and estimators are validated in a closed-loop fashion against the original "high fidelity" model. However, within the PAROC framework, the "high fidelity" modeling and analysis is performed in gPROMS ModelBuilder® while the model reduction as well as the formulation and solution of the multi-parametric programming problem is carried out in MATLAB. Thus currently, the closed-loop validation of the developed controller is done in MATLAB using the gPROMS ModelBuilder® tool gO:MATLAB. While this is a valid way of performing closed-loop validation, this does not allow for the use of the tools available in gPROMS (e.g. dynamic optimization). In Figure 12.3 a schematic that depicts the interaction between the different software packages is demonstrated. The actions enclosed in the gray rectangle happen in MATLAB environment. While the procedure in extracting the "high fidelity" model from gPROMS to a MATLAB friendly format is simple, straightforward, and thoroughly described in the former's user guides, the design of the controller in the POP mp-P software as well as interfacing the multi-parametric solution to the modeling software via gO:MATLAB requires a good knowledge

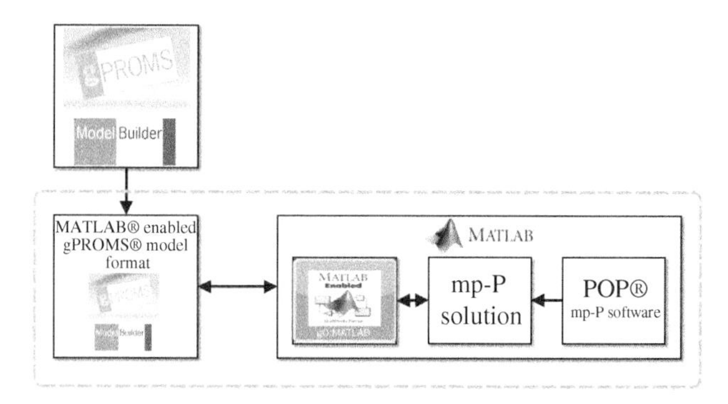

Figure 12.3 Software interactions for the implementation of the closed-loop validation through gO:MATLAB.

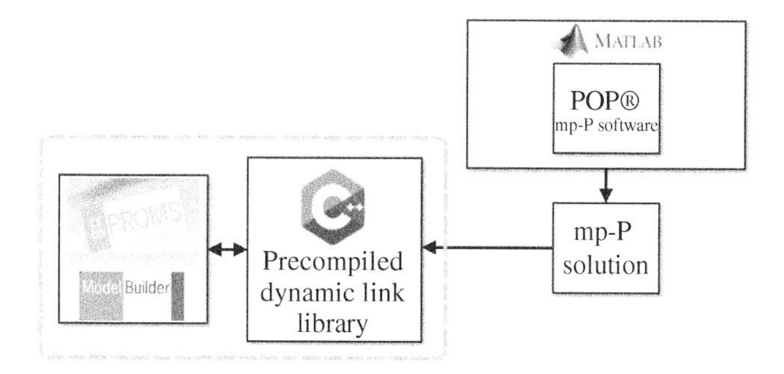

Figure 12.4 Software interactions for the implementation of the closed-loop validation through C++.

of both software packages. More specifically, the user needs to specify the information that is exchanged (including uncertain parameters, input actions, outputs and output set-points) in both ends. In addition, this procedure is conceptually problematic, as it suggests the test of a controller given a certain system rather than the test of an mp-MPC controlled system.

Therefore, a software solution that enables the direct export of the mp-MPC controller developed in MATLAB into gPROMS ModelBuilder as a foreign object has been developed. This foreign object, written in C++, loads the matrix representation and provides a simple look-up table as part of the gPROMS ModelBuilder architecture, similarly to, e.g. a Proportional-Integral-Derivative (PID) controller. Similarly to the gO:MATLAB case, a schematic is provided (Figure 12.4). In this case, the actions within the gray rectangle happen within the gPROMS platform.

12.3 Case Study: Distillation Column

The application of the PAROC Framework on a simple distillation column is the objective of this section. A distillation column is used to demonstrate the application of model reduction techniques on multi-parametric programming algorithms when systems of high dimensionality are considered. The simplified model for the binary separation of benzene and toluene is used as the basis for the design of an mp-MHE and an mp-MPC. Although the "high fidelity" model is relatively simple in nature (i.e. linear ordinary differential equation [ODEs] with nonlinear equilibrium relations), the use of approximation techniques is necessary for order reduction purposes. The estimator and controller are tested simultaneously against the "high fidelity" model in a closed loop formulation.

12.3.1 "High Fidelity" Modeling

The distillation column of this example has been modeled following the principles described in [48]. More specifically, the 32 tray distillation model consists of 32 state equations, 32 equilibrium relations, and 3 correlations of the volumetric flows. A

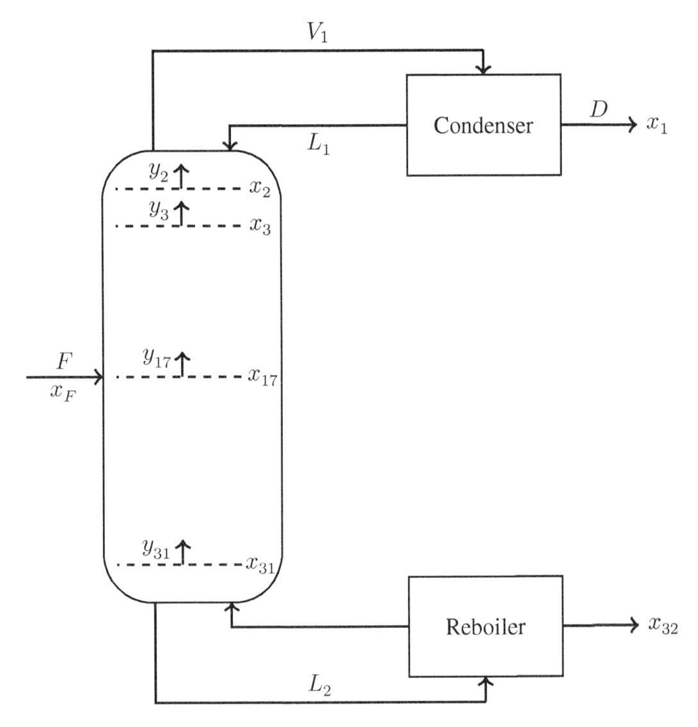

Figure 12.5 Graphical representation of the binary distillation column example.

graphical representation of the distillation column is presented in Figure 12.5.

It is assumed that in each tray equilibrium is established between the vapor phase and liquid phase composition. The condenser is considered to be a total condenser. The differential equations of the distillation column are categorized as follows (Eqs. (12.3a)–(12.3e)):

Condenser: $$\frac{dx_1}{dt} = \frac{1}{A_{cond}} V(y_2 - x_1), \tag{12.3a}$$

Trays 2–16: $$\frac{dx_i}{dt} = \frac{1}{A_{tray}} [L_1(x_{i-1} - x_i) - V(y_i - y_{i+1})], \tag{12.3b}$$

Feed Tray: $$\frac{dx_{17}}{dt} = \frac{1}{A_{tray}} [Fx_F + L_1x_{16} - L_2x_{17} - V(y_{17} - y_{18})], \tag{12.3c}$$

Trays 18–31: $$\frac{dx_i}{dt} = \frac{1}{A_{tray}} [L_2(x_{i-1} - x_i) - V_1(y_i - y_{i+1})], \tag{12.3d}$$

Reboiler: $$\frac{dx_{32}}{dt} = \frac{1}{A_{reb}} [L_2x_{31} - (F - D)(x_{32} - Vy_{32})], \tag{12.3e}$$

Equilibrium relation: $$\alpha = \frac{y_i(1 - x_i)}{x_i(1 - y_i)}, \quad V = L_1 + D, \tag{12.3f}$$

Flows correlation and ratios: $$L_2 = F + L_1, \quad RR = \frac{L_1}{D}. \tag{12.3g}$$

The algebraic equations that complete the "high fidelity" distillation column model are described by Eqs. (12.3f) and (12.3g).

The 32 states of the "high fidelity" model are defined by the liquid phase compositions. The condenser ratio (RR) is the only degree of freedom in the model. The composition of the liquid phase in the 32nd tray (i.e. the reboiler) is treated as the model output. For completeness purposes, the values of the model parameters are given in Table 12.5. The model is implemented in gPROMS.

Table 12.5 Parameter values for the distillation column model.

Symbol	Parameter value
F	0.4
D	0.2
A_{cond}	0.5
A_{tray}	0.25
A_{reb}	1.0
α	1.6

12.3.2 Model Approximation

Although this "high fidelity" model is a simple approach to distillation column modeling, the size of the model and its nonlinearity makes its use for optimization and control via multi-parametric programming a challenging issue. For that reason, model reduction techniques have been applied. More specifically, nonlinear balanced truncation, a snapshots based technique, is applied on the original model. According to this technique, we consider a nonlinear system of ODEs of the form of Eq. (12.4).

$$\begin{aligned} \dot{x}(t) &= f(x(t), u(t)), \\ y(t) &= h(x(t)). \end{aligned} \tag{12.4}$$

We derive a transformation matrix T in order to project the state vector x on a lower order subspace $\overline{x}$. The transformation matrix is based on empirical gramians or covariance matrices, which are computed via system simulation data [46, 49]. The transformed system of Eq. (12.5) is treated with linear balanced truncation methods in order to reduce the system to the controllable/essential states. The resulting 2-state model is presented in Eq. (12.6).

$$\begin{aligned} \dot{x}(t) &= Tf(T^{-1}\overline{x}(t), u(t)), \\ y(t) &= h(T^{-1}\overline{x}(t)), \end{aligned} \tag{12.5}$$

$$\begin{aligned} x_{T+1} &= \begin{bmatrix} 0.9546 & 0.05113 \\ -0.04809 & 0.3834 \end{bmatrix} x_T + \begin{bmatrix} -0.09323 \\ -0.0596 \end{bmatrix} u_T \\ &\quad + \begin{bmatrix} 0.0097686 \\ 0.045933 \end{bmatrix} w_T, \\ y_T &= \begin{bmatrix} -0.1009 & 0.06461 \end{bmatrix} x_T, \\ T_s &= 1s. \end{aligned} \tag{12.6}$$

In this case, y_T is the liquid phase composition of the reboiler, u_T is the condenser ratio, and w_T is considered to be a disturbance in the liquid composition of the feed that varies according to a Gaussian distribution. For the entire procedure of the reduction as well as the cross-validation of the approximate model against the original model, the reader is referred to [50].

The state-space model is used for the design of an mp-MHE as well as an mp-MPC, which are applied and tested against the original model.

12.3.3 Multi-parametric Programming, Control, and Estimation

This section focuses on the design of the estimator, the controller, and their solution via multi-parametric programming. This case study focuses on the synergetic effect of mp-MHE and mp-MPC on a reduced model against the dynamics of the original "high fidelity" system. The same state-space model computed in the previous section is used for the derivation of the mp-MHE and mp-MPC formulation. The objective of the mp-MHE formulation is (i) the estimation of the values of state vector and (ii) the estimation of the noise. The noise is not treated as uncertainty by the mp-MPC. The objective of the mp-MPC is to produce bottom product of a certain liquid composition regardless the noise in the inlet liquid composition, which varies for as much as 10% from the nominal value. The variables in the two multi-parametric problems have been normalized to vary between 0 and 1, while standard multi-parametric techniques were followed for the solution of the problems [51] in MATLAB, using the POP toolbox.

In the case of the mp-MHE, the 7 entries in the parameter vector consist of (i) the measured/calculated state vector, (ii) the previous

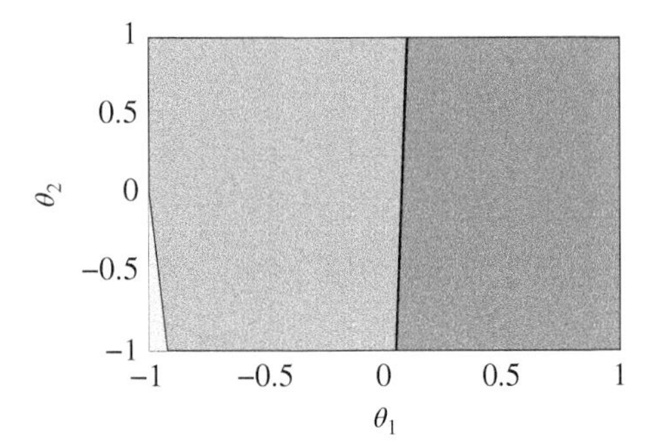

Figure 12.6 Critical regions for the solution of the mp-MHE problem of the binary distillation column. θ_1 and θ_2 correspond to estimated states x_1 and x_2, respectively. The rest of the parametric vector values have been fixed for plotting purposes.

and current measured outputs, and (iii) the previous control actions. In the case of the mp-MPC, the 4 entries in the parameter vector consist simply of (i) the estimated states, (ii) the current time output, and (iii) the output set-point as defined by the user. Figures 12.6 and 12.7 present the solution of the multi-parametric programming problems in the form of two-dimensional projections of the critical space. The projections are based on the states

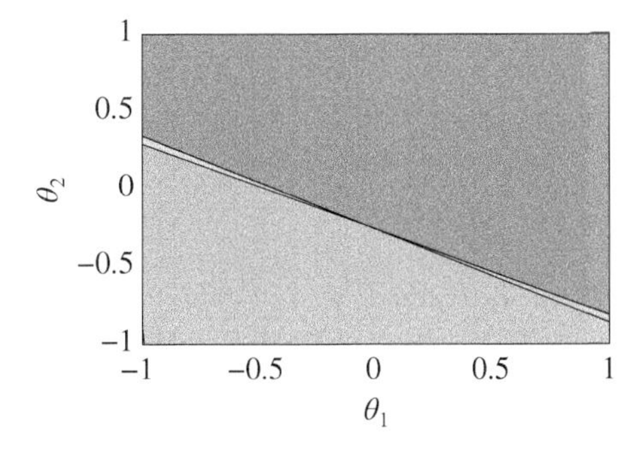

Figure 12.7 Critical regions for the solution of the mp-MPC problem of the binary distillation column. θ_1 and θ_2 correspond to estimated states x_1 and x_2, respectively. The rest of the parametric vector values have been fixed for plotting purposes.

variables of the parameter vectors, while the values of the rest of the parameters are set to certain values within their feasible bounds in order for the graphs to be generated.

12.3.4 Closed-Loop Validation

In this last step of the PAROC Framework, the testing of the controller and the estimator against the original "high fidelity" model developed in gPROMS takes place. The response and performance of the controller are evaluated as well as the ability of the mp-MHE to estimate the real states.

As in the previous case study, the controller scheme is tested against the "high fidelity" model through gO:MATLAB. The results of the closed loop validation are presented in Figure 12.8.

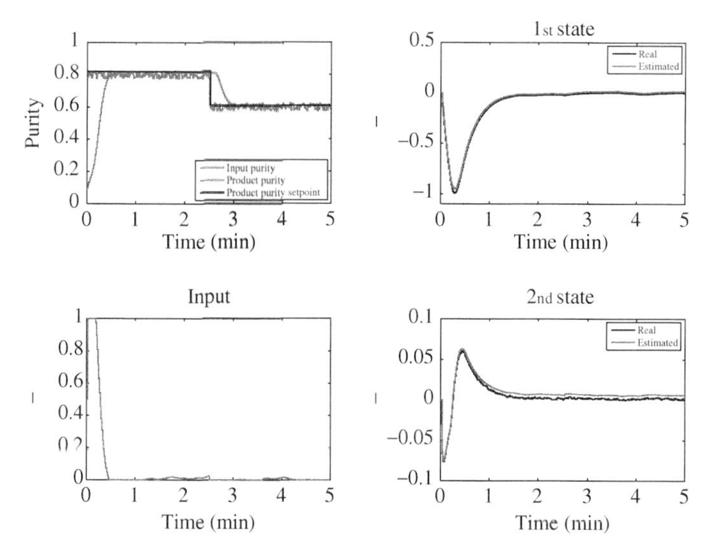

Figure 12.8 Closed loop validation results. Both the mp-MHE and mp-MPC schemes are tested against the "high fidelity" model of the binary distillation column. The desired purity is preserved regardless of the input purity variations, the approximate states are estimated effectively. Note that the lack of y-legend on the approximate state graphs reflects the lack of physical meaning of the approximate states.

The closed loop validation results show that both the estimator and the controller have been sufficiently well designed. More specifically the mp-MHE manages to accurately estimate the values of the states regardless the massive model reduction and the noise in the product purity inlet while the mp-MPC, using the estimated states, manages to accurately meet the set-point with a maximum 2.3% offset, after steady state has been reached.

It is worth noting the following:

- The open-loop simulations of the system determined the maximum feasible purity of the product, under steady state, for a condenser ration of 0. During the open-loop simulations, disturbance in the feed composition was not taken into account.
- The closed-loop simulations of Figure 12.8 feature two set-points. The set-points are the maximum feasible product purities, as determined by the aforementioned procedure, for nominal feed compositions of 0.8 and 0.6, respectively.
- It is observed that during the step change, the controller forces the condenser ratio to 0, regardless the estimation results, in order to reach the set-points as soon as possible, which is an expected behavior.

12.3.5 Conclusion

In this case study, a simple distillation column model was considered, consisting of 32 states. Nonlinear model reduction techniques were applied to reduce the order of the model and acquire a linear representation around a steady-state point. The linear state-space model was effectively used for the design of an mp-MHE as well as an mp-MPC. The two systems were used in parallel and tested against the original model. The results showed that the procedure followed was able to accurately drive the system into a user defined set-point as well as efficiently discard the noise in the composition of the liquid phase input. Similar distillation units have been extensively used to demonstrate the interactions between design optimization and control using novel algorithms on mixed-integer design optimization [52–56].

12.4 Case Study: Simple Buffer Tank

12.5 The Tank Example

This case study focuses on a simple tank. A sinusoidal inlet flow signal is introduced to a tank the outlet of which is manipulated via an mp-MPC. The purpose of the controller is to maintain a certain liquid volume within the tank regardless of the inlet. The sinusoidal form of the inlet is handled as a bounded parametric uncertainty via the control problem. The set-point is determined as a function of the nominal value and maximum deviation of the sinusoidal inlet flow rate. The tank volume is therefore inferred by that.

12.5.1 "High Fidelity" Dynamic Modeling

The model of the tank is presented in Eq. (12.7)

$$\begin{aligned}
\frac{dV}{dt} &= F_{in} - F_{out}, \\
F_{out} &= a \cdot V, \\
F_{in} &= F_{nom} + F_{dev} \cdot \sin\left(\frac{t}{\text{freq}}\right), \\
\text{freq} &= \frac{1}{2 \cdot \pi},
\end{aligned} \tag{12.7}$$

where V is the volume of the liquid within the tank, F_{in} and F_{out} are the inlet and outlet flow rate, respectively, a is a proportionality parameter and the control variable, F_{nom} and F_{dev} are the nominal inlet flow rate and its deviation, respectively, and *freq* is the sinusoidal signal frequency. Note that V is the state of the system and is bi-linear with the control variable a, which makes the linearization of the system necessary for multi-parametric programming. Alternatively, a robust reformulation of the system in discrete time would alleviate the need of approximation resulting in the consideration of the exact model in the multi-parametric programming formulation. In this context, F_{in} being treated as bounded parametric uncertainty does not interfere with the linearity of the state-space formulation.

12.5.2 Model Approximation

The approximation of the tank model in Eq. (12.7) takes place in the System Identification Toolbox in MATLAB and results into the linear state-space in Eq. (12.8).

$$\begin{aligned} x_{T+1} &= 0.9980 \cdot x_T - 0.2536 \cdot u_{c,T} + 0.1003 \cdot d_T, \\ y_T &= 109.9320 \cdot x_T - 2.0378 \cdot u_{c,T}, \\ T_s &= 0.01s, \end{aligned} \tag{12.8}$$

where x_T are the identified states, $u_{c,T}$ is the proportionality parameter, d_T is the flow at the inlet of the tank, and y is the liquid volume in the tank.

The step and impulse responses of the system are presented in Figure 12.9a, b, respectively.

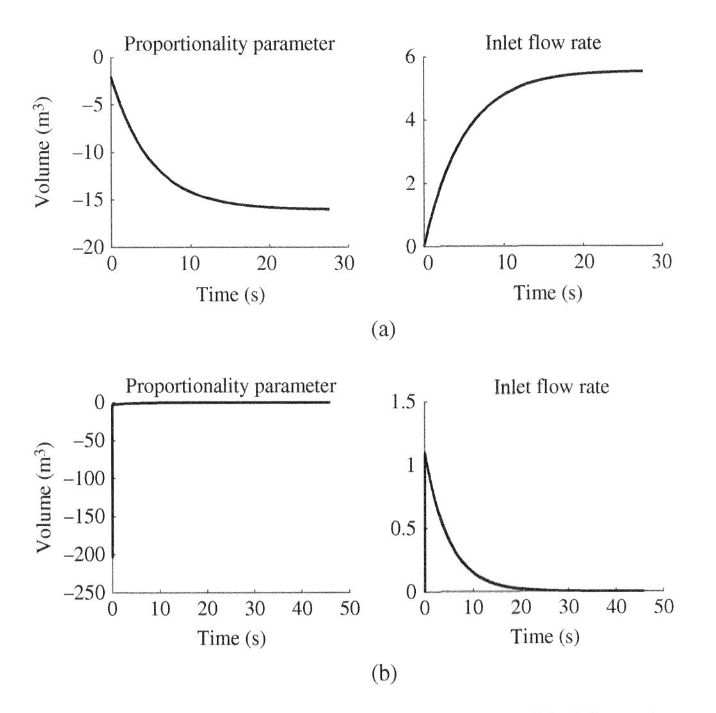

Figure 12.9 Step response (a) and impulse response (b) of the tank approximate model. The negative values on the y-axis are attributed to the $0m^3$ initial condition and valve opening that has a negative effected on the liquid volume in the tank.

12.5.3 Design of the Multi-parametric Model Predictive Controller

The mp-MPC problem is formulated and solved using POP where the optimal control action is generated as a map of solutions and as a function of the problem parameters (as shown in previous chapter). The problem formulation is based on Eq. (11.22) and the tuning of the controller is presented in Table 12.6.

The design variable corresponding to the volume of the tank is introduced as a parameter at the upper bound of the output of the system. Furthermore, the bounds for the disturbance of the system d_{min} and d_{max} have been chosen such that $\{F_{nom,min} - F_{dev,max}, \dots, F_{nom,max} + F_{dev,max}\} \subseteq \{d_{min}, \dots, d_{max}\}$. The purpose of the controller is to maintain the volume of the liquid within the tank at a certain set-point with a maximum deviation of less than 5%, effectively rejecting the disturbance introduced at the inlet of the tank.

Table 12.6 Weight tuning for the mp-MPC of the tank.

MPC design parameters	Value
N	10
M	1
$QR_k, \forall k \in \{1, \dots, N\}$	10
$R_k, \forall k \in \{1, \dots, M\}$	10^{-7}
x_{min}	-10^3
x_{max}	10^3
u_{min}	0
u_{max}	1
y_{min}	0
y_{max}	10
d_{min}	0
d_{max}	5

12.5.4 Closed-Loop Validation

The validation step is presented in Figure 12.10 where the controller is tested against the original high fidelity model. The closed loop validation of the controller happens for the following process characteristics:

- Nominal inlet flow rate: 1.5 m^3/s
- Inlet flow rate deviation: 0.5 m^3/s
- Volume set-point: 2 m^3

Note that in the case of the tank the set-point for the volume V^{SP} is dynamically defined as the maximum value of the inlet flow rate (i.e. if $F_{in} = F_{nom} + F_{dev} \cdot \sin(t/freq)$ [m^3/s] then $V^{SP} = F_{in} + F_{out}$ [m^3]).

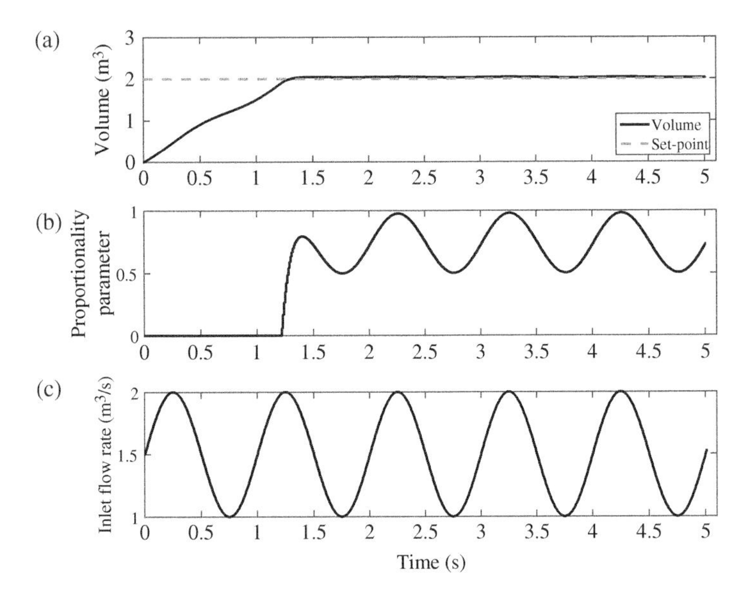

Figure 12.10 Closed-loop validation results for liquid volume set-point tracking. (a) Liquid volume (system output) set-point tracking; (b) corresponding proportionality parameter setting (optimal action); (c) inlet flow rate sinusoidal fluctuation (system disturbance).

12.5.5 Conclusion

In this case study, a buffer tank was considered the volume of which was considered as uncertain. Data-based system identification techniques were applied to acquire a linear representation of the dynamic model. The linear state-space model was effectively used for the design of an mp-MPC. The controller was used and tested against the original model. The results showed that the procedure followed was able to accurately drive the system into a user defined set-point as well as efficiently discard the inlet disturbance.

12.6 Concluding Remarks

In this chapter the PAROC framework was used to decompose the problem into a series of steps. In the first step, a "high fidelity" model of the process is developed. In the second step, the initial "high fidelity" model is reduced using system identification as well as recently developed model-reduction techniques. This reduced model is used in the next step to obtain the explicit control law of the system, which is calculated once and offline. This control law is then validated in silico against the original "high fidelity" model, thus closing the loop.

The key advantages of PAROC are (i) its effortless applicability on fundamentally different classes of problems, (ii) its decomposed nature which allows for different advanced applications in the different steps, (iii) the inherently offline solution of the problem thus limiting the online computational burden, and (iv) the validation of the exact solution derived from the simplified model against the "high fidelity" model through the interconnectivity of the different software packages.

References

1 Aspen Tech (2001–2016) Aspen plus. www.aspentech.com/products/aspen-plus.aspx, 2016.
2 Process Systems Enterprise (1997–2019) gPROMS. URL http://www.psenterprise.com/gproms/, 2016.

3 Lee, H.H., Koppel, L.B., and Lim, H.C. (1972) Integrated approach to design and control of a class of countercurrent processes. *Industrial & Engineering Chemistry Process Design and Development*, 11 (3), 376–382.
4 Narraway, L.T., Perkins, J.D., and Barton, G.W. (1991) Interaction between process design and process control: economic analysis of process dynamics. *Journal of Process Control*, 1 (5), 243–250.
5 Luyben, M.L. and Floudas, C.A. (1994) Analyzing the interaction of design and control—1. A multiobjective framework and application to binary distillation synthesis. *Computers and Chemical Engineering*, 18 (10), 933–969, doi: 10.1016/0098-1354(94)E0013-D.
6 Mohideen, M.J., Perkins, J.D., and Pistikopoulos, E.N. (1996) Optimal synthesis and design of dynamic systems under uncertainty. *Computers and Chemical Engineering*, 20, Supplement 2, S895–S900, doi: 10.1016/0098-1354(96)00157-3.
7 Van Schijndel, J.M.G. and Pistikopoulos, E.N. (2000) Towards the integration of process design, process control and process operability - current stares and furore trends. *Foundations of Computer-Aided Process Design*, 96, 99–112.
8 Sakizlis, V., Perkins, J.D., and Pistikopoulos, E.N. (2004) Recent advances in optimization-based simultaneous process and control design. *Computers and Chemical Engineering*, 28 (10), 2069–2086, doi: 10.1016/j.compchemeng.2004.03.018.
9 Flores-Tlacuahuac, A. and Biegler, L.T. (2007) Simultaneous mixed-integer dynamic optimization for integrated design and control. *Computers and Chemical Engineering*, 31 (5–6), 588–600, doi: 10.1016/j.compchemeng.2006.08.010.
10 Würth, L., Hannemann, R., and Marquardt, W. (2011) A two-layer architecture for economically optimal process control and operation. *Journal of Process Control*, 21 (3), 311–321, doi: 10.1016/j.jprocont.2010.12.008.
11 Yuan, Z., Chen, B., Sin, G., and Gani, R. (2012) State-of-the-art and progress in the optimization-based simultaneous design and control for chemical processes. *AIChE Journal*, 58 (6), 1640–1659, doi: 10.1002/aic.13786.

12 Diangelakis, N.A., Manthanwar, A.M., and Pistikopoulos, E.N. (2014) A framework for design and control optimisation: application on a CHP system, in *Proceedings of the 8th International Conference on Foundations of Computer-Aided Process Design, Computer aided chemical engineering*, vol. 34 (eds M.R. Eden, J.D. Siirola, and G.P. Towler), Elsevier, pp. 765–770, doi: 10.1016/B978-0-444-63433-7.50112-7. URL http://www.sciencedirect.com/science/article/pii/B9780444634337501127.
13 Liu, P., Georgiadis, M.C., and Pistikopoulos, E.N. (2013) An energy systems engineering approach for the design and operation of microgrids in residential applications. *Chemical Engineering Research and Design*, 91 (10), 2054–2069. The 60th Anniversary of the European Federation of Chemical Engineering (EFCE).
14 Sakizlis, V., Perkins, J.D., and Pistikopoulos, E.N. (2003) Parametric controllers in simultaneous process and control design optimization. *Industrial and Engineering Chemistry Research*, 42 (20), 4545–4563.
15 Shobrys, D. and White, D. (2002) Planning, scheduling and control systems: why cannot they work together. *Computers and Chemical Engineering*, 26 (2), 149–160.
16 Mahadevan, R., Doyle, F. III, and Allcock, A. (2002) Control-relevant scheduling of polymer grade transitions. *AIChE Journal*, 48 (8), 1754–1764, doi: 10.1002/aic.690480816.
17 Chatzidoukas, C., Perkins, J.D., Pistikopoulos, E.N., and Kiparissides, C. (2003) Optimal grade transition and selection of closed-loop controllers in a gas-phase olefin polymerization fluidized bed reactor. *Chemical Engineering Science*, 58 (16), 3643–3658.
18 Chatzidoukas, C., Kiparissides, C., Perkins, J.D., and Pistikopoulos, E.N. (2003) Optimal grade transition campaign scheduling in a gas-phase polyolefin FBR using mixed integer dynamic optimization. *Computer Aided Chemical Engineering*, 15 (C), 744–747.
19 Nyström, R., Franke, R., Harjunkoski, I., and Kroll, A. (2005) Production campaign planning including grade transition sequencing and dynamic optimization. *Computers and Chemical Engineering*, 29 (10), 2163–2179.

20 Flores-Tlacuahuac, A. and Grossmann, I.E. (2006) Simultaneous cyclic scheduling and control of a multiproduct CSTR. *Industrial and Engineering Chemistry Research*, 45 (20), 6698–6712.

21 Flores-Tlacuahuac, A. and Grossmann, I.E. (2006) An effective MIDO approach for the simultaneous cyclic scheduling and control of polymer grade transition operations. *Computer Aided Chemical Engineering*, 21 (C), 1221–1226, doi: 10.1016/S1570-7946(06)80213-0.

22 Harjunkoski, I., Nyström, R., and Horch, A. (2009) Integration of scheduling and control-theory or practice? *Computers and Chemical Engineering*, 33 (12), 1909–1918, doi: 10.1016/j.compchemeng.2009.06.016.

23 Biegler, L. and Zavala, V. (2009) Large-scale nonlinear programming using IPOPT: an integrating framework for enterprise-wide dynamic optimization. *Computers and Chemical Engineering*, 33 (3), 575–582.

24 Subramanian, K., Rawlings, J.B., Maravelias, C.T., Flores-Cerrillo, J., and Megan, L. (2013) Integration of control theory and scheduling methods for supply chain management. *Computers and Chemical Engineering*, 51, 4–20, doi: 10.1016/j.compchemeng.2012.06.012.

25 Subramanian, K., Maravelias, C.T., and Rawlings, J.B. (2012) A state-space model for chemical production scheduling. *Computers and Chemical Engineering*, 47, 97–110, doi: 10.1016/j.compchemeng.2012.06.025. URL http://www.sciencedirect.com/science/article/pii/S0098135412002098.

26 Zhuge, J. and Ierapetritou, M.G. (2014) Integration of scheduling and control for batch processes using multi-parametric model predictive control. *AIChE Journal*, 60 (9), 3169–3183, doi: 10.1002/aic.14509. URL http://dx.doi.org/10.1002/aic.14509.

27 Kopanos, G.M. and Pistikopoulos, E.N. (2014) Reactive scheduling by a multiparametric programming rolling horizon framework: a case of a network of combined heat and power units. *Industrial and Engineering Chemistry Research*, 53 (11), 4366–4386, doi: 10.1021/ie402393s.

28 Baldea, M. and Harjunkoski, I. (2014) Integrated production scheduling and process control: a systematic review. *Computers and Chemical Engineering*, 71, 377–390.

29 You, F. and Grossmann, I.E. (2008) Design of responsive supply chains under demand uncertainty. *Computers and Chemical Engineering*, 32 (12), 3090–3111.

30 Khajuria, H. and Pistikopoulos, E.N. (2011) Dynamic modeling and explicit/multi-parametric MPC control of pressure swing adsorption systems. *Journal of Process Control*, 21 (1), 151–163, doi: 10.1016/j.jprocont.2010.10.021. URL http://www.sciencedirect.com/science/article/pii/S0959152410002167.

31 Krieger, A., Panoskaltsis, N., Mantalaris, A., Georgiadis, M.C., and Pistikopoulos, E.N. (2014) Modeling and analysis of individualized pharmacokinetics and pharmacodynamics for volatile anesthesia. *IEEE Transactions on Biomedical Engineering*, 61 (1), 25–34, doi: 10.1109/TBME.2013.2274816.

32 Saltelli, A., Chan, K., and Scott, E.M. (2000) *Sensitivity analysis, Wiley series in probability and statistics*, John Wiley & Sons, New York, Chichester, and Weinheim, etc.

33 Kontoravdi, C., Asprey, S.P., Pistikopoulos, E.N., and Mantalaris, A. (2005) Application of global sensitivity analysis to determine goals for design of experiments: an example study on antibody-producing cell cultures. *Biotechnology Progress*, 21 (4), 1128–1135, doi: 10.1021/bp050028k. URL http://dx.doi.org/10.1021/bp050028k.

34 Kiparissides, A., Kucherenko, S.S., Mantalaris, A., and Pistikopoulos, E.N. (2009) Global sensitivity analysis challenges in biological systems modeling. *Industrial and Engineering Chemistry Research*, 48 (15), 7168–7180, doi: 10.1021/ie900139x. URL http://dx.doi.org/10.1021/ie900139x.

35 Nascu, I., Oberdieck, R., and Pistikopoulos, E.N. (2016) Explicit hybrid model predictive control strategies for intravenous anaesthesia. *Computers and Chemical Engineering*, doi: 10.1016/j.compchemeng.2017.01.033.

36 Lambert, R.S., Rivotti, P., and Pistikopoulos, E.N. (2013) A Monte-Carlo based model approximation technique for linear model predictive control of nonlinear systems. *Computers and Chemical Engineering*, 54, 60–67, doi: 10.1016/j.compchemeng.2013.03.004. URL http://www.sciencedirect.com/science/article/pii/S0098135413000690.

37 Van Overschee, P. and De Moor, B. (1994) N4SID: subspace algorithms for the identification of combined deterministic-stochastic systems. *Automatica*, 30 (1), 75–93, doi: 10.1016/0005-1098(94)90230-5.

38 Ljung, L. (1987) *System Identification: Theory for the User*, Prentice-Hall, Inc., River, NJ, USA, 0138816409.

39 Boukouvala, F. and Floudas, C.A. (2016) ARGONAUT: AlgoRithms for global optimization of coNstrAined grey-box compUTational problems. *Optimization Letters*, 11, 1–19, doi: 10.1007/s11590-016-1028-2.

40 Boukouvala, F., Misener, R., and Floudas, C.A. (2016) Global optimization advances in mixed-integer nonlinear programming, MINLP, and constrained derivative-free optimization, CDFO. *European Journal of Operational Research*, 252 (3), 701–727, doi: 10.1016/j.ejor.2015.12.018.

41 Kouramas, K. and Pistikopoulos, E.N. (2014) *Wind turbines modeling and control*, vol. 5–7, Wiley-VHC Verlag GmbH & Co. KGaA, doi: 10.1002/9783527631209.ch47.

42 Sakizlis, V., Perkins, J.D., and Pistikopoulos, E.N. (2005) Explicit solutions to optimal control problems for constrained continuous-time linear systems. *IEEE Proceedings: Control Theory and Applications*, 152 (4), 443–452, doi: 10.1049/ip-cta:20059041.

43 Sun, M., Chachuat, B., and Pistikopoulos, E.N. (2016) Design of multi-parametric NCO tracking controllers for linear dynamic systems. *Computers and Chemical Engineering*, 92, 64–77, doi: 10.1016/j.compchemeng.2016.04.038.

44 Narciso, D.A. and Pistikopoulos, E.N. (2008) A combined balanced truncation and multi-parametric programming approach for linear model predictive control, in *18th European Symposium on Computer Aided Process Engineering, Computer Aided Chemical Engineering*, vol. 25 (eds B. Braunschweig and X. Joulia), Elsevier, pp. 405–410, doi: 10.1016/S1570-7946(08)80072-7. URL http://www.sciencedirect.com/science/article/pii/S1570794608800727.

45 Rivotti, P., Lambert, R.S., and Pistikopoulos, E.N. (2012) Combined model approximation techniques and multiparametric programming for explicit nonlinear model predictive control. *Computers and Chemical Engineering*, 42, 277–287, doi: 10.1016/j.compchemeng.2012.01.009. URL http://www.sciencedirect.com/science/article/pii/S0098135412000191.

46 Hahn, J. and Edgar, T.F. (2002) An improved method for nonlinear model reduction using balancing of empirical gramians. *Computers and Chemical Engineering*, 26 (10), 1379–1397, doi: 10.1016/S0098-1354(02)00120-5. URL http://www.sciencedirect.com/science/article/pii/S0098135402001205.

47 Pistikopoulos, E.N. (2009) Perspectives in multiparametric programming and explicit model predictive control. *AIChE Journal*, 55 (8), 1918–1925, doi: 10.1002/aic.11965. URL http://dx.doi.org/10.1002/aic.11965.

48 Benallou, A., Seborg, D., and Mellichamp, D. (1986) Dynamic compartmental models for separation processes. *AIChE Journal*, 32 (7), 1067–1078.

49 Hahn, J., Edgar, T., and Marquardt, W. (2003) Controllability and observability covariance matrices for the analysis and order reduction of stable nonlinear systems. *Journal of Process Control*, 13 (2), 115–127.

50 Lambert, R., Nascu, I., and Pistikopoulos, E.N. (2013) Simultaneous reduced order multi-parametric moving horizon estimation and model predictive control, in *Dynamics and Control of Process Systems*, Elsevier, IFAC, IFAC proceedings volumes, pp. 45–50, doi: 10.3182/20131218-3-IN-2045.00071.

51 Bemporad, A., Morari, M., Dua, V., and Pistikopoulos, E.N. (2002) The explicit linear quadratic regulator for constrained systems. *Automatica*, 38 (1), 3–20, doi: 10.1016/S0005-1098(01)00174-1. URL http://www.sciencedirect.com/science/article/pii/S0005109801001741.

52 Pistikopoulos, E.N. and Sakizlis, V. (2002) Simultaneous design and control optimization under uncertainty in reaction/separation systems. *AIChE Symposium Series*, 98 (326), 223–238.

53 Bansal, V., Ross, R., Perkins, J.D., Pistikopoulos, E.N., and de Wolf, S. (2002) An industrial case study in simultaneous design and control using mixed-integer dynamic optimization. *Computer Aided Chemical Engineering*, 10 (C), 163–168.

54 Bansal, V., Perkins, J.D., and Pistikopoulos, E.N. (2002) A case study in simultaneous design and control using rigorous, mixed-integer dynamic optimization models. *Industrial and Engineering Chemistry Research*, 41 (4), 760–778, doi: 10.1021/ie010156n.

55 Ross, R., Bansal, V., Perkins, J.D., Pistikopoulos, E.N., Koot, G.L.M., and van Schijndel, J.M.G. (1999) Optimal design and control of an industrial distillation system. *Computers and Chemical Engineering*, 23, S875–S878, doi: 10.1016/S0098-1354(99)80215-4.

56 Bansal, V., Perkins, J.D., Pistikopoulos, E.N., Ross, R., and van Schijndel, J.M.G. (2000) Simultaneous design and control optimisation under uncertainty. *Computers and Chemical Engineering*, 24 (2–7), 261–266, doi: 10.1016/S0098-1354(00)00475-0.

A

Appendix for the mp-MPC Chapter 10

In the following text, the matrices that define the multi-parametric quadratic programming (mp-QP) problem stemming from the multi-parametric model predictive control (mp-MPC) of Eq. (10.30) are presented. The resulting mp-QP follows the form of Eq. (A.1) and that u is the vector of optimization variables and x_0 the vector of unknown but bounded parameters:

$$\begin{aligned} \underset{u}{\text{minimize}} \quad & u^T H u + u^T Z x_0 \\ \text{subject to} \quad & Gu \leq W + S x_0 \\ & CR_A x_0 \leq CR_b. \end{aligned} \tag{A.1}$$

Multi-parametric Optimization and Control, First Edition.
Efstratios N. Pistikopoulos, Nikolaos A. Diangelakis, and Richard Oberdieck.

$$
G = \begin{bmatrix}
-6 & -5 & -4 & -3 & -2 & -1 & 0 & 0 & 0 & 0 \\
-5 & -4 & -3 & -2 & -1 & 0 & 0 & 0 & 0 & 0 \\
-4 & -3 & -2 & -1 & 0 & 0 & 0 & 0 & 0 & 0 \\
-3 & -2 & -1 & 0 & 0 & 0 & 0 & 0 & 0 & 0 \\
-2 & -1 & 0 & 0 & 0 & 0 & 0 & 0 & 0 & 0 \\
-1 & 0 & 0 & 0 & 0 & 0 & 0 & 0 & 0 & 0 \\
-1 & 0 & 0 & 0 & 0 & 0 & 0 & 0 & 0 & 0 \\
0 & -1 & 0 & 0 & 0 & 0 & 0 & 0 & 0 & 0 \\
0 & 0 & -1 & 0 & 0 & 0 & 0 & 0 & 0 & 0 \\
0 & 0 & 0 & -1 & 0 & 0 & 0 & 0 & 0 & 0 \\
0 & 0 & 0 & 0 & -1 & 0 & 0 & 0 & 0 & 0 \\
0 & 0 & 0 & 0 & 0 & -1 & 0 & 0 & 0 & 0 \\
0 & 0 & 0 & 0 & 0 & 0 & -1 & 0 & 0 & 0 \\
0 & 0 & 0 & 0 & 0 & 0 & 0 & -1 & 0 & 0 \\
0 & 0 & 0 & 0 & 0 & 0 & 0 & 0 & -1 & 0 \\
0 & 0 & 0 & 0 & 0 & 0 & 0 & 0 & 0 & -1 \\
0 & 0 & 0 & 0 & 0 & 0 & 0 & 0 & 0 & 0 \\
0 & 0 & 0 & 0 & 0 & 0 & 0 & 0 & 0 & 0 \\
0 & 0 & 0 & 0 & 0 & 0 & 0 & 0 & 0 & 1 \\
0 & 0 & 0 & 0 & 0 & 0 & 0 & 0 & 1 & 0 \\
0 & 0 & 0 & 0 & 0 & 0 & 0 & 1 & 0 & 0 \\
0 & 0 & 0 & 0 & 0 & 0 & 1 & 0 & 0 & 0 \\
0 & 0 & 0 & 0 & 0 & 1 & 0 & 0 & 0 & 0 \\
0 & 0 & 0 & 0 & 1 & 0 & 0 & 0 & 0 & 0 \\
0 & 0 & 0 & 1 & 0 & 0 & 0 & 0 & 0 & 0 \\
0 & 0 & 1 & 0 & 0 & 0 & 0 & 0 & 0 & 0 \\
0 & 1 & 0 & 0 & 0 & 0 & 0 & 0 & 0 & 0 \\
1 & 0 & 0 & 0 & 0 & 0 & 0 & 0 & 0 & 0 \\
1 & 0 & 0 & 0 & 0 & 0 & 0 & 0 & 0 & 0 \\
2 & 1 & 0 & 0 & 0 & 0 & 0 & 0 & 0 & 0 \\
3 & 2 & 1 & 0 & 0 & 0 & 0 & 0 & 0 & 0 \\
4 & 3 & 2 & 1 & 0 & 0 & 0 & 0 & 0 & 0 \\
5 & 4 & 3 & 2 & 1 & 0 & 0 & 0 & 0 & 0 \\
6 & 5 & 4 & 3 & 2 & 1 & 0 & 0 & 0 & 0 \\
7.13 & 6.52 & 5.91 & 5.29 & 4.68 & 4.06 & 3.45 & 2.84 & 2.22 & 1.61 \\
-3.74 & -3.36 & -2.99 & -2.61 & -2.24 & -1.87 & -1.49 & -1.12 & -0.74 & -0.37 \\
3.74 & 3.36 & 2.99 & 2.61 & 2.24 & 1.87 & 1.49 & 1.12 & 0.74 & 0.37 \\
-7.13 & -6.52 & -5.91 & -5.29 & -4.68 & -4.06 & -3.45 & -2.84 & -2.22 & -1.61
\end{bmatrix} \tag{A.2}
$$

$$W = \begin{bmatrix} 10 \\ 10 \\ 10 \\ 10 \\ 10 \\ 1 \\ 10 \\ 1 \\ 1 \\ 1 \\ 1 \\ 1 \\ 1 \\ 1 \\ 1 \\ 1 \\ 10 \\ 10 \\ 1 \\ 1 \\ 1 \\ 1 \\ 1 \\ 1 \\ 1 \\ 1 \\ 1 \\ 10 \\ 1 \\ 10 \\ 10 \\ 10 \\ 10 \\ 10 \\ 1 \\ 1 \\ 1 \\ 1 \end{bmatrix}, \quad \text{and } S = \begin{bmatrix} 1 & 7 \\ 1 & 6 \\ 1 & 5 \\ 1 & 4 \\ 1 & 3 \\ 0 & 0 \\ 1 & 2 \\ 0 & 0 \\ 0 & 0 \\ 0 & 0 \\ 0 & 0 \\ 0 & 0 \\ 0 & 0 \\ 0 & 0 \\ 0 & 0 \\ 0 & 0 \\ -1 & -1 \\ 1 & 1 \\ 0 & 0 \\ 0 & 0 \\ 0 & 0 \\ 0 & 0 \\ 0 & 0 \\ 0 & 0 \\ 0 & 0 \\ 0 & 0 \\ 0 & 0 \\ -1 & -2 \\ 0 & 0 \\ -1 & -3 \\ -1 & -4 \\ -1 & -5 \\ -1 & -6 \\ -1 & -7 \\ -0.61 & -7.75 \\ 0.37 & 4.11 \\ -0.37 & -4.11 \\ 0.61 & 7.75 \end{bmatrix} \tag{A.3}$$

$$CR_A = \begin{bmatrix} -1 & 0 \\ 0 & -1 \\ 1 & 0 \\ 0 & 1 \end{bmatrix}, \quad \text{and } CR_b = \begin{bmatrix} 10 \\ 10 \\ 10 \\ 10 \end{bmatrix} \tag{A.4}$$

$$H = \begin{bmatrix} 457.49 & 395.24 & 334 & 274.76 & 218.52 & 166.28 & 119.04 & 77.8 & 43.55 & 17.31 \\ 395.24 & 344.63 & 293.01 & 242.39 & 193.77 & 148.15 & 106.54 & 69.92 & 39.3 & 15.68 \\ 334 & 293.01 & 252.02 & 210.02 & 169.02 & 130.03 & 94.04 & 62.04 & 35.05 & 14.05 \\ 274.76 & 242.39 & 210.02 & 177.66 & 144.28 & 111.91 & 81.54 & 54.17 & 30.79 & 12.42 \\ 218.52 & 193.77 & 169.02 & 144.28 & 119.54 & 93.78 & 69.04 & 46.29 & 26.54 & 10.79 \\ 166.28 & 148.15 & 130.03 & 111.91 & 93.78 & 75.67 & 56.54 & 38.41 & 22.29 & 9.16 \\ 119.04 & 106.54 & 94.04 & 81.54 & 69.04 & 56.54 & 44.05 & 30.54 & 18.03 & 7.53 \\ 77.8 & 69.92 & 62.04 & 54.17 & 46.29 & 38.41 & 30.54 & 22.67 & 13.78 & 5.91 \\ 43.55 & 39.3 & 35.05 & 30.79 & 26.54 & 22.29 & 18.03 & 13.78 & 9.54 & 4.28 \\ 17.31 & 15.68 & 14.05 & 12.42 & 10.79 & 9.16 & 7.53 & 5.91 & 4.28 & 2.66 \end{bmatrix} \tag{A.5}$$

$$Z = \begin{bmatrix} 122.48 & 1037.45 \\ 101.24 & 891.72 \\ 81.99 & 749.99 \\ 64.74 & 614.26 \\ 49.49 & 486.53 \\ 36.25 & 368.8 \\ 25 & 263.07 \\ 15.75 & 171.34 \\ 8.51 & 95.61 \\ 3.26 & 37.88 \end{bmatrix} \tag{A.6}$$

B

Appendix for the mp-MPC Chapter 11

B.1 Matrices for the mp-QP Problem Corresponding to the Example of Section 11.3.2

In the following text, the matrices that define the multi-parametric quadratic programming (mp-QP) problem stemming from the multi-parametric model predictive control (mp-MPC) of Eq. (11.30) are presented. The resulting mp-QP follows the form of Eq. (B.1) and that u is the vector of optimization variables and $\theta = \left[y^{ref}\ x_0^1\ x_0^2\ u_{-1}\right]^T$ the vector of unknown but bounded parameters:

$$
\begin{aligned}
\underset{u}{\text{minimize}}\ & u^T H u + u^T Z \theta \\
\text{subject to}\ & Gu \le W + S\theta \\
& CR_A \theta \le CR_b.
\end{aligned}
\tag{B.1}
$$

Multi-parametric Optimization and Control, First Edition.
Efstratios N. Pistikopoulos, Nikolaos A. Diangelakis, and Richard Oberdieck.

$$
G = \begin{bmatrix}
0.00 & 0.00 & 0.00 & 0.00 & 0.00 & 0.00 & 0.00 & 0.00 & 0.00 & 0.00 \\
0.71 & 0.00 & 0.00 & 0.00 & 0.00 & 0.00 & 0.00 & 0.00 & 0.00 & 0.00 \\
0.00 & 0.00 & 0.00 & 0.00 & 0.00 & 0.00 & 0.00 & 0.00 & 0.00 & 0.00 \\
-0.71 & 0.00 & 0.00 & 0.00 & 0.00 & 0.00 & 0.00 & 0.00 & 0.00 & 0.00 \\
0.47 & 0.00 & 0.00 & 0.00 & 0.00 & 0.00 & 0.00 & 0.00 & 0.00 & 0.00 \\
-0.47 & 0.00 & 0.00 & 0.00 & 0.00 & 0.00 & 0.00 & 0.00 & 0.00 & 0.00 \\
0.58 & 0.58 & 0.00 & 0.00 & 0.00 & 0.00 & 0.00 & 0.00 & 0.00 & 0.00 \\
-0.58 & -0.58 & 0.00 & 0.00 & 0.00 & 0.00 & 0.00 & 0.00 & 0.00 & 0.00 \\
0.59 & 0.24 & 0.00 & 0.00 & 0.00 & 0.00 & 0.00 & 0.00 & 0.00 & 0.00 \\
-0.59 & -0.24 & 0.00 & 0.00 & 0.00 & 0.00 & 0.00 & 0.00 & 0.00 & 0.00 \\
0.50 & 0.50 & 0.50 & 0.00 & 0.00 & 0.00 & 0.00 & 0.00 & 0.00 & 0.00 \\
-0.50 & -0.50 & -0.50 & 0.00 & 0.00 & 0.00 & 0.00 & 0.00 & 0.00 & 0.00 \\
0.61 & 0.34 & 0.14 & 0.00 & 0.00 & 0.00 & 0.00 & 0.00 & 0.00 & 0.00 \\
-0.61 & -0.34 & -0.14 & 0.00 & 0.00 & 0.00 & 0.00 & 0.00 & 0.00 & 0.00 \\
0.45 & 0.45 & 0.45 & 0.45 & 0.00 & 0.00 & 0.00 & 0.00 & 0.00 & 0.00 \\
-0.45 & -0.45 & -0.45 & -0.45 & 0.00 & 0.00 & 0.00 & 0.00 & 0.00 & 0.00 \\
0.60 & 0.39 & 0.22 & 0.09 & 0.00 & 0.00 & 0.00 & 0.00 & 0.00 & 0.00 \\
-0.60 & -0.39 & -0.22 & -0.09 & 0.00 & 0.00 & 0.00 & 0.00 & 0.00 & 0.00 \\
0.41 & 0.41 & 0.41 & 0.41 & 0.41 & 0.00 & 0.00 & 0.00 & 0.00 & 0.00 \\
-0.41 & -0.41 & -0.41 & -0.41 & -0.41 & 0.00 & 0.00 & 0.00 & 0.00 & 0.00 \\
0.59 & 0.41 & 0.26 & 0.15 & 0.06 & 0.00 & 0.00 & 0.00 & 0.00 & 0.00 \\
-0.59 & -0.41 & -0.26 & -0.15 & -0.06 & 0.00 & 0.00 & 0.00 & 0.00 & 0.00 \\
0.38 & 0.38 & 0.38 & 0.38 & 0.38 & 0.38 & 0.00 & 0.00 & 0.00 & 0.00 \\
-0.38 & -0.38 & -0.38 & -0.38 & -0.38 & -0.38 & 0.00 & 0.00 & 0.00 & 0.00 \\
0.57 & 0.42 & 0.30 & 0.19 & 0.11 & 0.04 & 0.00 & 0.00 & 0.00 & 0.00 \\
-0.57 & -0.42 & -0.30 & -0.19 & -0.11 & -0.04 & 0.00 & 0.00 & 0.00 & 0.00 \\
0.35 & 0.35 & 0.35 & 0.35 & 0.35 & 0.35 & 0.35 & 0.00 & 0.00 & 0.00 \\
-0.35 & -0.35 & -0.35 & -0.35 & -0.35 & -0.35 & -0.35 & 0.00 & 0.00 & 0.00 \\
0.56 & 0.43 & 0.32 & 0.22 & 0.14 & 0.08 & 0.03 & 0.00 & 0.00 & 0.00 \\
-0.56 & -0.43 & -0.32 & -0.22 & -0.14 & -0.08 & -0.03 & 0.00 & 0.00 & 0.00 \\
0.33 & 0.33 & 0.33 & 0.33 & 0.33 & 0.33 & 0.33 & 0.33 & 0.00 & 0.00 \\
-0.33 & -0.33 & -0.33 & -0.33 & -0.33 & -0.33 & -0.33 & -0.33 & 0.00 & 0.00 \\
0.54 & 0.43 & 0.33 & 0.25 & 0.17 & 0.11 & 0.06 & 0.02 & 0.00 & 0.00 \\
-0.54 & -0.43 & -0.33 & -0.25 & -0.17 & -0.11 & -0.06 & -0.02 & 0.00 & 0.00 \\
0.32 & 0.32 & 0.32 & 0.32 & 0.32 & 0.32 & 0.32 & 0.32 & 0.32 & 0.00 \\
-0.32 & -0.32 & -0.32 & -0.32 & -0.32 & -0.32 & -0.32 & -0.32 & -0.32 & 0.00 \\
0.53 & 0.43 & 0.34 & 0.26 & 0.19 & 0.14 & 0.09 & 0.05 & 0.02 & 0.00 \\
-0.53 & -0.43 & -0.34 & -0.26 & -0.19 & -0.14 & -0.09 & -0.05 & -0.02 & 0.00 \\
0.30 & 0.30 & 0.30 & 0.30 & 0.30 & 0.30 & 0.30 & 0.30 & 0.30 & 0.30 \\
-0.30 & -0.30 & -0.30 & -0.30 & -0.30 & -0.30 & -0.30 & -0.30 & -0.30 & -0.30 \\
1.00 & 0.00 & 0.00 & 0.00 & 0.00 & 0.00 & 0.00 & 0.00 & 0.00 & 0.00 \\
-1.00 & 0.00 & 0.00 & 0.00 & 0.00 & 0.00 & 0.00 & 0.00 & 0.00 & 0.00 \\
-0.51 & -0.43 & -0.35 & -0.28 & -0.21 & -0.16 & -0.11 & -0.07 & -0.04 & -0.01 \\
0.51 & 0.43 & 0.35 & 0.28 & 0.21 & 0.16 & 0.11 & 0.07 & 0.04 & 0.01 \\
-0.53 & -0.43 & -0.34 & -0.26 & -0.20 & -0.14 & -0.09 & -0.05 & -0.03 & -0.01 \\
0.53 & 0.43 & 0.34 & 0.26 & 0.20 & 0.14 & 0.09 & 0.05 & 0.03 & 0.01 \\
0.53 & 0.43 & 0.34 & 0.26 & 0.20 & 0.14 & 0.09 & 0.05 & 0.02 & 0.01 \\
-0.53 & -0.43 & -0.34 & -0.26 & -0.20 & -0.14 & -0.09 & -0.05 & -0.02 & -0.01
\end{bmatrix}
$$

$$
= \begin{bmatrix}
-0.52 & -0.43 & -0.35 & -0.27 & -0.21 & -0.15 & -0.11 & -0.07 & -0.04 & -0.01 \\
0.52 & 0.43 & 0.35 & 0.27 & 0.21 & 0.15 & 0.11 & 0.07 & 0.04 & 0.01 \\
0.52 & 0.42 & 0.34 & 0.26 & 0.19 & 0.14 & 0.09 & 0.05 & 0.03 & 0.01 \\
-0.52 & -0.42 & -0.34 & -0.26 & -0.19 & -0.14 & -0.09 & -0.05 & -0.03 & -0.01 \\
0.45 & 0.38 & 0.31 & 0.24 & 0.19 & 0.14 & 0.10 & 0.06 & 0.03 & 0.01 \\
-0.45 & -0.38 & -0.31 & -0.24 & -0.19 & -0.14 & -0.10 & -0.06 & -0.03 & -0.01 \\
-0.51 & -0.41 & -0.33 & -0.26 & -0.19 & -0.14 & -0.09 & -0.05 & -0.03 & -0.01 \\
0.51 & 0.41 & 0.33 & 0.26 & 0.19 & 0.14 & 0.09 & 0.05 & 0.03 & 0.01 \\
-0.15 & -0.13 & -0.10 & -0.08 & -0.07 & -0.05 & -0.03 & -0.02 & -0.01 & -0.01 \\
0.15 & 0.13 & 0.10 & 0.08 & 0.07 & 0.05 & 0.03 & 0.02 & 0.01 & 0.01 \\
0.17 & 0.14 & 0.11 & 0.09 & 0.06 & 0.05 & 0.03 & 0.02 & 0.01 & 0.00 \\
-0.17 & -0.14 & -0.11 & -0.09 & -0.06 & -0.05 & -0.03 & -0.02 & -0.01 & 0.00 \\
0.01 & 0.01 & 0.01 & 0.01 & 0.01 & 0.01 & 0.00 & 0.00 & 0.00 & 0.00 \\
-0.01 & -0.01 & -0.01 & -0.01 & -0.01 & -0.01 & 0.00 & 0.00 & 0.00 & 0.00 \\
-0.03 & -0.03 & -0.02 & -0.02 & -0.01 & -0.01 & -0.01 & 0.00 & 0.00 & 0.00 \\
0.03 & 0.03 & 0.02 & 0.02 & 0.01 & 0.01 & 0.01 & 0.00 & 0.00 & 0.00
\end{bmatrix}
\tag{B.2}
$$

$$
W = \begin{bmatrix}
7.07 \\ 0.71 \\ 7.07 \\ 0.71 \\ 0.71 \\ 0.71 \\ 0.58 \\ 0.58 \\ 0.36 \\ 0.36 \\ 0.50 \\ 0.50 \\ 0.20 \\ 0.20 \\ 0.45 \\ 0.45 \\ 0.13 \\ 0.13 \\ 0.41 \\ 0.41 \\ 0.09 \\ 0.09 \\ 0.38 \\ 0.38
\end{bmatrix}, \text{and } S = \begin{bmatrix}
0.00 & -0.71 & -0.71 & 0.00 \\
0.00 & 0.00 & 0.00 & -0.71 \\
0.00 & 0.71 & 0.71 & 0.00 \\
0.00 & 0.00 & 0.00 & 0.71 \\
0.00 & -0.24 & -0.71 & -0.47 \\
0.00 & 0.24 & 0.71 & 0.47 \\
0.00 & 0.00 & 0.00 & -0.58 \\
0.00 & 0.00 & 0.00 & 0.58 \\
0.00 & -0.12 & -0.47 & -0.59 \\
0.00 & 0.12 & 0.47 & 0.59 \\
0.00 & 0.00 & 0.00 & -0.50 \\
0.00 & 0.00 & 0.00 & 0.50 \\
0.00 & -0.07 & -0.34 & -0.61 \\
0.00 & 0.07 & 0.34 & 0.61 \\
0.00 & 0.00 & 0.00 & -0.45 \\
0.00 & 0.00 & 0.00 & 0.45 \\
0.00 & -0.04 & -0.26 & -0.60 \\
0.00 & 0.04 & 0.26 & 0.60 \\
0.00 & 0.00 & 0.00 & -0.41 \\
0.00 & 0.00 & 0.00 & 0.41 \\
0.00 & -0.03 & -0.21 & -0.59 \\
0.00 & 0.03 & 0.21 & 0.59 \\
0.00 & 0.00 & 0.00 & -0.38 \\
0.00 & 0.00 & 0.00 & 0.38
\end{bmatrix}
$$

$$
= \begin{bmatrix} 0.06 \\ 0.06 \\ 0.35 \\ 0.35 \\ 0.05 \\ 0.05 \\ 0.33 \\ 0.33 \\ 0.04 \\ 0.04 \\ 0.32 \\ 0.32 \\ 0.03 \\ 0.03 \\ 0.30 \\ 0.30 \\ 3.00 \\ 3.00 \\ 0.01 \\ 0.01 \\ 0.03 \\ 0.03 \\ 0.01 \\ 0.01 \\ 0.11 \\ 0.11 \\ 0.18 \\ 0.18 \\ 0.80 \\ 0.80 \\ 1.02 \\ 1.02 \\ 2.27 \\ 2.27 \\ 2.04 \\ 2.04 \\ 2.26 \\ 2.26 \\ 2.28 \\ 2.28 \end{bmatrix}, \text{and } S = \begin{bmatrix} 0.00 & -0.02 & -0.17 & -0.57 \\ 0.00 & 0.02 & 0.17 & 0.57 \\ 0.00 & 0.00 & 0.00 & -0.35 \\ 0.00 & 0.00 & 0.00 & 0.35 \\ 0.00 & -0.02 & -0.14 & -0.56 \\ 0.00 & 0.02 & 0.14 & 0.56 \\ 0.00 & 0.00 & 0.00 & -0.33 \\ 0.00 & 0.00 & 0.00 & 0.33 \\ 0.00 & -0.01 & -0.12 & -0.54 \\ 0.00 & 0.01 & 0.12 & 0.54 \\ 0.00 & 0.00 & 0.00 & -0.32 \\ 0.00 & 0.00 & 0.00 & 0.32 \\ 0.00 & -0.01 & -0.11 & -0.53 \\ 0.00 & 0.01 & 0.11 & 0.53 \\ 0.00 & 0.00 & 0.00 & -0.30 \\ 0.00 & 0.00 & 0.00 & 0.30 \\ 0.00 & 0.00 & 0.00 & 0.00 \\ 0.00 & 0.00 & 0.00 & 0.00 \\ 0.00 & 0.01 & 0.10 & 0.51 \\ 0.00 & -0.01 & -0.10 & -0.51 \\ -0.01 & 0.01 & 0.11 & 0.53 \\ 0.01 & -0.01 & -0.11 & -0.53 \\ 0.01 & -0.01 & -0.11 & -0.53 \\ -0.01 & 0.01 & 0.11 & 0.53 \\ -0.02 & 0.01 & 0.10 & 0.52 \\ 0.02 & -0.01 & -0.10 & -0.52 \\ -0.03 & -0.01 & -0.10 & -0.52 \\ 0.03 & 0.01 & 0.10 & 0.52 \\ -0.13 & -0.01 & -0.08 & -0.45 \\ 0.13 & 0.01 & 0.08 & 0.45 \\ -0.18 & 0.01 & 0.10 & 0.51 \\ 0.18 & -0.01 & -0.10 & -0.51 \\ -0.38 & 0.00 & 0.03 & 0.15 \\ 0.38 & 0.00 & -0.03 & -0.15 \\ -0.34 & 0.00 & -0.03 & -0.17 \\ 0.34 & 0.00 & 0.03 & 0.17 \\ -0.38 & 0.00 & 0.00 & -0.01 \\ 0.38 & 0.00 & 0.00 & 0.01 \\ -0.38 & 0.00 & 0.01 & 0.03 \\ 0.38 & 0.00 & -0.01 & -0.03 \end{bmatrix} \tag{B.3}
$$

$$CR_A = \begin{bmatrix} 1 & -2 & 1 & -1 \\ 0 & 1 & 0 & 0 \\ 0 & 0 & 1 & 0 \\ 0 & 0 & 0 & 1 \\ -1 & 2 & -1 & 1 \\ 0 & -1 & 0 & 0 \\ 0 & 0 & -1 & 0 \\ 0 & 0 & 0 & -1 \\ 0 & 2 & 1 & 0 \\ 0 & 1 & 2 & 0 \\ 0 & -2 & -1 & 0 \\ 0 & -1 & -2 & 0 \end{bmatrix}, \text{and } CR_b = \begin{bmatrix} 50 \\ 10 \\ 10 \\ 1 \\ 50 \\ 10 \\ 10 \\ 1 \\ 25 \\ 3 \\ 25 \\ 3 \end{bmatrix} \tag{B.4}$$

$$H = \begin{bmatrix} 30326.79 & 23769.55 & 18068.84 & 13219.56 & 9208.71 & 6011.29 & 3586.29 \\ 23769.55 & 18701.98 & 14276.01 & 10490.92 & 7341.63 & 4815.13 & 2886.42 \\ 18068.84 & 14276.01 & 10948.96 & 8087.40 & 5690.63 & 3753.55 & 2263.16 \\ 13219.56 & 10490.92 & 8087.40 & 6009.10 & 4255.72 & 2826.56 & 1716.53 \\ 9208.71 & 7341.63 & 5690.63 & 4255.72 & 3037.00 & 2034.16 & 1246.51 \\ 6011.29 & 4815.13 & 3753.55 & 2826.56 & 2034.16 & 1376.45 & 853.12 \\ 3586.29 & 2886.42 & 2263.16 & 1716.53 & 1246.51 & 853.12 & 536.45 \\ 1872.73 & 1514.50 & 1194.46 & 912.61 & 668.95 & 463.48 & 296.20 \\ 785.59 & 638.37 & 506.44 & 389.81 & 288.47 & 202.42 & 131.67 \\ 211.88 & 173.03 & 138.12 & 107.13 & 80.08 & 56.95 & 37.76 \end{bmatrix}$$

$$= \begin{bmatrix} 1872.73 & 785.59 & 211.88 \\ 1514.50 & 638.37 & 173.03 \\ 1194.46 & 506.44 & 138.12 \\ 912.61 & 389.81 & 107.13 \\ 668.95 & 288.47 & 80.08 \\ 463.48 & 202.42 & 56.95 \\ 296.20 & 131.67 & 37.76 \\ 167.21 & 76.21 & 22.50 \\ 76.21 & 36.14 & 11.16 \\ 22.50 & 11.16 & 3.86 \end{bmatrix} \tag{B.5}$$

$$Z = \begin{bmatrix} -857.43 & 1714.86 & 14829.13 & 60653.38 \\ -641.79 & 1283.58 & 11418.92 & 47539.10 \\ -465.69 & 931.38 & 8517.05 & 36137.69 \\ -325.12 & 650.24 & 6107.52 & 26439.13 \\ -216.09 & 432.17 & 4166.34 & 18417.43 \\ -134.59 & 269.17 & 2661.50 & 12022.58 \\ -76.62 & 153.24 & 1553.00 & 7172.59 \\ -38.19 & 76.38 & 792.84 & 3745.46 \\ -15.29 & 30.59 & 325.02 & 1571.18 \\ -3.93 & 7.86 & 85.55 & 423.75 \\ & & & \end{bmatrix} \tag{B.6}$$

Index

Multi-parametric Optimization and Control, First Edition.
Efstratios N. Pistikopoulos, Nikolaos A. Diangelakis, and Richard Oberdieck.